Role of Interfaces in Concrete

Proceedings of the International Seminar
held at the University of Dundee, Scotland, UK
on 7 September 1999

Edited by

Ravindra K. Dhir
Director, Concrete Technology Unit
University of Dundee

and

Peter C. Hewlett
Director, British Board of Agrément

ThomasTelford

INTRODUCTION

Given its composite and heterogeneous nature, it can be argued that the interfaces in concrete define its overall performance from strength to durability. Interfaces occur between all phases of concrete, at the aggregate/matrix bond, between fine aggregate and paste and matrix and reinforcement. Furthermore, a whole area of cement and concrete science can be devoted to the understanding of the interfacial effects on surface impregnation and coating technology, screeding, topping and repair systems.

Within the interfacial zone, the concrete performance may be quite different to that of the bulk. Understanding the characteristics of the interfacial zones and how they are affected by mix constituents and production variables is a key to improving the performance of concrete. Unfortunately, these interfaces are complex structures that are affected by the wide range of intrinsic and extrinsic factors.

The inclusion of pozzolanic cements such as PFA, GGBS, metakaolin and microsilica can greatly improve the paste microstructure, reducing its porosity and improving the paste/aggregate bond. It has also been found using computer models that groups of cements and filler can be mixed to minimise both aggregate-size and capillary-size voids thereby maximising the solids content. These optimised mixes make best use of the available binder materials but there is still much to be established fundamentally.

In addition, not all processes occurring at interfaces are physical. There may be a concentration of fluid and ionic transportation in these zones and, at present, how to optimise the interfaces to minimise, for example, chloride ingress and carbonation are only empirically understood.

The reduction in free water content and water/cement ratio, as well as the use of polymers, generally brings about an improvement in interfacial bond quality. Developments in both admixture and polymer technology are allowing both very high strength concrete and high durability concrete to be produced and placed on-site.

It has also become increasingly evident that the interfacial zone between the reinforcement and paste plays an important role in both the passivation and corrosion processes as well as influencing structural behaviour in terms of initial bond slip and load to first crack.

The Proceedings for the Seminar; *Role of Interfaces in Concrete* dealt with two main themes, namely (i) Basic Mechanisms and (ii) Achieving Performance. Each of these themes was opened by a Leader Paper presented by an expert in that field. There were a total of 22 papers presented during the Seminar which have been compiled into these Proceedings.

Dundee September 1999

Ravindra K Dhir Peter C Hewlett

ORGANISING COMMITTEE
Concrete Technology Unit

Professor R K Dhir, OBE (Chairman)

Dr M R Jones (Secretary)

Mr M D Newlands (Joint Secretary)

Professor P C Hewlett
British Board of Agrément

Dr N A Henderson
Mott MacDonald Ltd

Professor V K Rigopoulou
National Technical University of Athens, Greece

Dr S Y N Chan
Hong Kong Polytechnic University

Dr N Y Ho
L & M Structural Systems, Singapore

Dr M J McCarthy

Dr M C Limbachiya

Dr T D Dyer

Dr K A Paine

Dr T G Jappy

Mr P A J Tittle

Mr J C Knights

Mr S R Scott (Unit Assistant)

Miss A M Duncan (Unit Secretary)

NATIONAL TECHNICAL COMMITTEE

Mr P Barber
Manager of the Scheme, The Quality Scheme for Ready Mixed Concrete

Professor A W Beeby
Professor of Structural Design, University of Leeds

Mr B V Brown
Divisional Technical Executive, Readymix (UK) Ltd.

Dr T W Broyd
Technology Development Director, W S Atkins Ltd.

Professor J H Bungey
Professor of Civil Engineering, University of Liverpool

Dr P S Chana
Director, CRIC, Imperial College of Science, Technology & Medicine

Professor J L Clarke
Principal Engineer, The Concrete Society

Dr P C Das
Group Manager, Structures Management, Highways Agency

Dr S B Desai, OBE
Principal Civil Engineer, Department of the Environment, Transport and the Regions

Professor R K Dhir, OBE (Chairman)
Director, Concrete Technology Unit, University of Dundee

Mr C R Ecob
Director Special Services Division, Mott MacDonald Ltd.

Professor F P Glasser
University of Aberdeen

Professor T A Harrison
Technical Consultant, Quarry Products Association

Professor P C Hewlett
Director, British Board of Agrément

Professor J Innes
Director of Roads, Scottish Office

NATIONAL TECHNICAL COMMITTEE (CONTINUED)

Mr K A L Johnson
Director, AMEC Civil Engineering Ltd.

Dr M R Jones
Senior Lecturer, Concrete Technology Unit, University of Dundee

Mr P Livesey
National Technical Services Manager, Castle Cement Ltd.

Professor A E Long
Director of School, Queens University of Belfast

Professor P S Mangat
Head of Research, Sheffield Hallam University

Mr G Masterton
Director, Babtie Group Ltd.

Professor G C Mays
Director of Civil Engineering, Cranfield University

Mr L H McCurrich
Technology Development Consultant, Fosroc Construction

Professor R S Narayanan
Partner, SB Tietz & Partners Consulting Engineers

Dr P J Nixon
Head, Centre for Concrete Construction, Building Research Establishment Ltd.

Dr W F Price
Senior Associate, Messrs Sandberg

Professor G Somerville, OBE
Director of Engineering, British Cement Association

Professor D C Spooner
Director, Materials and Standards, British Cement Association

Dr H P J Taylor
Director, Tarmac Precast Concrete Ltd.

Mr M Walker
Technical Manager, The Concrete Society

Dr R J Woodward
Senior Project Manager, Transport Research Laboratory

SPONSORING ORGANISATIONS WITH EXHIBITION (CONTINUED)

Readymix (UK) Ltd.

Rugby Cement

Scottish Enterprise Tayside

Sika Ltd.

SKW - MBT Construction Chemicals

Thomas Telford Publishing Ltd.

United Kingdom Quality Ash Association

W A Fairhurst & Partners

ADDITIONAL EXHIBITORS

Christison Scientific Equipment Ltd.

CMS Pozament Limited

The Concrete Society

David Ball Group plc.

E & FN Spon

Flexcrete Ltd.

Germann Instruments A/S, Denmark

Natural Cement Distribution Limited

Palladian Publications Ltd.

Quality Scheme for Ready Mixed Concrete

UK Certification Authority for Reinforcing Steel

Wacker-Chemie GmbH, Germany

Wexham Developments

CONTENTS

Preface	iii
Introduction	iv
Organising Committee	v
International Advisory Committee	vi
National Technical Committee	viii
Supporting Institutions	x
Sponsoring Organisations With Exhibition	xi
Additional Exhibitors	xii

Leader Paper
Research directions for lifetime prediction and long-term durability   1
T Igusa and S P Shah, Northwestern University, United States of America

THEME 1 BASIC MECHANISMS

Keynote Paper
Why are aggregate/matrix interfaces that thick?   17
J M J M Bijen, INTRON B.V. and M R de Rooij, Delft University of Technology, Netherlands

Corrosion and bond characteristics of reinforcing steel in normal Portland and fly ash concretes   27
L Amleh and M S Mirza

Polymer - concrete interactions   39
N R Short and I M Shaw

Unsaturated mass transfer within hardened cement paste modified by the partial replacement with silica fume   51
D Fairhurst and A K Platten

Hygric material properties of mechanically loaded concrete   63
J Drchalová, S Hošková, J Toman, T Klečka, and R Černý

Toughening mechanisms and fracture characteristics of concrete   71
V Bílek and Z Keršner

Study on influences of surface microcracks on permeability and frost resistance of steam cured concrete products   79
M Aba, M Shoya and K Otsuka

LEADER PAPER

INTRODUCTION

Concrete technology has advanced dramatically in many fronts during the past few decades. Water-reducing admixtures make it possible to easily produce concrete with strengths over 100 MPa. Sophisticated structural analysis methods have facilitated the analysis of concrete structures with complex configurations and loads. Modern construction processes have reduced costs and increased quality of concrete structures. Thus, concrete has been, and continues to be, the material of choice for most of the world's transportation infrastructure.

While progress is continuing in all phases of concrete technology, the steady and highly visible deterioration of infrastructure has brought widespread scientific and engineering focus on long-term durability and service life prediction. The main issues in concrete durability are related to chemical and physical attack, either on the exposed surface or within the concrete material. Chemical attack has a variety of forms, including chloride ion permeation and the resulting corrosion of reinforcing steel, sulfate attack, carbonation, alkali-aggregate reaction, and leaching and efflorescence. Physical attack includes freeze-thaw cycles, wetting and drying cycles, wear and abrasion and temperature fluctuations [1]. The resulting deterioration, unfortunately, also has a variety of forms, including cracking, spalling, and loss of reinforcement bond.

These forms of attack and their influence on durability are highly interrelated, making research on concrete durability a challenging task. Herein, a discussion of some of the important issues on service life prediction with emphasis on concrete durability is presented. To limit the scope, the discussion does not go beyond the three major forms of degradation: chloride permeability, freeze-thaw cycles, and reinforcing steel corrosion.

RESEARCH OBJECTIVES

Research Focus

A major research goal is to predict the long-term performance of a concrete system using data collected during the material design and placement phases, with focus on the interactions between chloride permeability, freeze-thaw, and corrosion. The scientific challenge is significant since expected lifetimes of 50, 100, or more years far exceeds the short-term time available for materials testing. Furthermore, it is necessary to go beyond established long-term reliability techniques which are based on universally applicable extreme-value theories. Herein, two basic research areas are addressed: the development of destructive and non-destructive accelerated testing methods and the development of models based on materials and statistical sciences.

Materials-Based Durability Models

There has been a major research effort at the NSF Center for Advanced Cement-Based Materials (ACBM) and at other research institutions in the development of analytical and computational models for cementitious materials. These models describe the multiple length scales in concrete micro- and meso-structure, and include processes such as the transport mechanisms for chloride ions and other aggressive chemicals.

For cementitious materials applications, the system loads contain wide variability and extremes – thus future performance must be expressed in statistical terms. There are statistical techniques in "reliability engineering" to extrapolate long-term performance predictions from short-term data [2]. However, most of these techniques are generic and work only for systems with available long-term performance data. Thus, the theory would not be applicable to new testing methods. The view herein is that a synergy of materials and statistical sciences is the best approach for analyzing the durability of complex materials such as concrete. Materials science would be used to identify key observables that can be used as indicators of durability. These observables can be of a variety of forms, including direct measurements such as temperature and chemical concentrations, and cause-and-effect measures such as the behavior under short-term loads. Emphasis on durability will be on deterioration and other degradation mechanisms from the micron to the centimeter length scales. The materials-based reliability science would guide and be guided by innovative experimentation described in the next subsection.

Science-Based Short-Term Tests

We need to go beyond established reliability techniques based on accelerated testing [2] to extrapolate durability from short-term data. The usual tests involve relatively high-speed repetition of exaggerated magnitude loads such as temperature fluctuations, stresses, or chemical gradients and concentrations. The results of such tests would be used to predict the behavior of the structure under similar loads at much longer time scales and lower magnitudes. In the materials-based durability science, the various components of loads such as stress, surface attack, freeze-thaw cycles, and chemical attack would be integrated in deterioration models with multi-scale dimensions. With this scientific view of the cementitious material, an innovative battery of short-term tests would be developed based not only on duplicating long-term loads at accelerated pace and magnitude levels, but also on the integrated relationships between observables and performance. Destructive tests would be used for test specimens extracted during concrete placement. Non-destructive tests would be needed to experimentally investigate the relations between observables and deterioration for *in situ* specimens. Both types of tests are also needed to verify, calibrate, and guide the development of the materials-based durability models.

The observables of interest would be very broadly defined in materials science terms. It may be that an important observable may be one that has not yet been experimentally measured with precision. A current set of examples of such observables is associated with the micron-scale interface transition zone surrounding aggregate surfaces. The challenge here is to explore innovative experimental methods at various length scales to measure the critical observables. While the needs here are driven by the materials science – it may also be that advanced sensors developed for other applications may yield important observables in cementitious materials. In either case, the experimental data obtained from such sensors would be used in statistical methods to bring new insights in the materials-based durability science.

Impact on Materials Design

The testing methods and the approach to reliability analysis would be applicable to new designs of cementitious materials. A successful design of cementitious composites with targeted properties requires understanding and control of its inherently variable structure over a wide range of length scales.

The chloride permeability of concrete is generally evaluated by direct measurement of the diffusion coefficient, by monitoring chloride profiles moving into a concrete section, or by indirect methods such as the electrical conductivity test AASHTO T277, commonly known as the Rapid Chloride Permeability Test. Since, direct diffusivity methods take a long time for completion, indirect accelerated methods such as AASHTO T277 which measure ionic conduction under the application of electric field are used. The relationships between the direct and indirect tests would have to be experimentally calibrated, and further investigated by analytical and numerical models.

As for physical damage caused by freeze-thaw loads or reinforcement corrosion, indirect non-destructive testing methods will be needed. Damage may take the form of deteriorated mechanical properties, discrete defects such as cracks, regions of damaged material, or reduced bonding between any interfaces. The use of ultrasonic and sonic stress wave-based methods enables deep penetration and offers direct information concerning the elastic moduli, or other constitutive parameters, and presence of flaws. However, the current state of the art is still in the developmental phase. Advances have to be made to generate suitably controlled signals in concrete, to effectively detect signals, and to interpret quantitative measurements for evaluation of various aspects of degradation [22, 23, 24, 25, 26].

An important part of the research work is the development of new equipment and suitable techniques for specific application to concrete system testing. An effective use of ultrasound/sound in this regard will require the following: a device to generate repeatable and controllable ultrasound/sound signals in concrete structures; a technique for far-side (bottom) crack detection in pavements; and a technique for characterization of interfacial properties. The stress wave source will most likely be mechanical, as opposed to piezoelectric, in nature. A pneumatic source or an electric impact gun may provide repeatable and controllable signals. Emphasis will need to be placed on tailoring the sources to the various developed non-destructive stress wave techniques. Thus, the signal source will have to be variable in terms of frequency content and generated directivity field. The source must also be applicable to relatively rough surfaces, such as those found on unprepared airport and highway pavements.

Modeling Program – Direct and Indirect Methods

Ionic conduction through concrete depends on its microstructure (10^{-6} m) and mesostructure (10^{-3} m). Conductive microstructure includes gel and capillary pores in cement paste, voids in the weak matrix-aggregate interfacial zone and microcracks due to volumetric changes. Mesostructure, which includes aggregates, air voids, and macrocracks, also influences conductivity by providing obstacles to the macro flow of ions. Hence, conductivity in concrete is characterized by wide range of feature sizes, from nanometer pores to centimeter size aggregates. For any study of transport properties of concrete to be valid, conductivity of samples should be large compared to the length scale (correlation length) of the heterogeneities so that average flow rates can be used to analyze the system [27].

The conductivity of porous media, such as concrete, has traditionally been described using continuum and discrete models. Continuum modeling is the classical engineering approach for describing complex materials like concrete, characterized by several length scales. Effective properties of a porous medium are defined as the averages of corresponding microscopic quantities.

The averages must be taken over a volume small enough compared to the volume of the system, but large enough for applicable equations of change (e.g., Darcy's law and Fick's Law) to hold when applied to that volume. Predictions are difficult due to the complex nature of the system (length scales and interconnectivity) and many empirical factors are often added to account for disparities between theoretical and experimental results. Discrete models on the other hand do not suffer from these limitations because the description can be anywhere from sub-microscopic level to megascopic levels. These models are based on a discretized description of the material in terms of networks. The only difficulty is the amount of computational effort required and knowledge of applicable mathematical equations. Some of the ideas applied included percolation theory, diffusion processes, fractal concepts and universal scaling laws, which describe how and under what conditions the effective macroscopic properties of the system may be independent of its microscopic details [27].

The effective medium theory models are continuum models based on the assumption that effective transport properties of a disordered medium can be replaced by a hypothetical homogeneous one which mimics the behavior of the disordered one and then calculate the homogeneous medium's properties. In case such a substitution is possible, the problem of calculating the effective transport properties of the homogeneous medium is much simpler than the original problem as this is only an approximation of the original problem. Historically, there have been two approaches to this problem. The *t-matrix* and *non-self-consistent approximation* is applicable in the case when isolated inclusions are embedded in a continuous matrix consisting of a single phase and the properties of the effective medium are obtained by placing a sphere (or an ellipse) of the homogeneous or the effective medium in this matrix [28]. In the second approach [29], each homogeneity is embedded in the effective medium itself, the unknown properties of which are determined in such a way that the volume average over all the inhomogeneities yields no extra fields in the medium. This is usually called the effective-medium theory (EMT) and is the focus of this paper. Garboczi, et al., [30] provide a comprehensive account on the application of the theory to cementitious materials.

There exists a vast literature on predicting the effective conductivity of the two-phase composite materials [31, 32]. Virtually all of these models use perfect interface between the constituents, which is not the case in reality and especially in concrete. Recently, Torquato & Rintoul [33] used classical minimum energy principles to find very sharp, rigorous bounds on the effective properties of a class of composites with imperfect interfaces that incorporate crucial microstructural information at the interface. Cheng & Torquato [34] have proposed an extension of the Rayleigh's dipole model to a two-phase media with a super thin interface of different properties than other two phases. It is difficult to apply such models to concrete because of the randomness of the material and difficulty of computation. The EMT models are very suitable to the conduction problem in concrete because of its basis in electric fields and averaging approximations involved.

The history of the application of the EMT in concrete is very short. Garboczi and Bentz [35] were the first to apply EMT to mortars in calculating the effective conductivity of a three phase composite consisting of aggregate, interface transition zone (ITZ) and matrix. They developed formulations for self-consistent (SC-EMT) and differential (D-EMT) methods. The SC-EMT has roots going back to Bruggeman [29]. In the D-EMT the exact solution for very low volume concentration can be used to generate an approximate equation that can be solved for the effective conductivity.

The negative sign is due to the dilution effect, where increased aggregate volume fraction decreases the paste volume fraction resulting in reduced conductivity. The magnitude of m is larger than unity because of the tortuosity effect, where the tortuous paths of the ion flow leads to further reduction in conductivity.

Garboczi, et al., [30] generalized the dilute limit equation in two ways: they considered multiple aggregate sizes (grades) and the effect of a conductive ITZ. Their result is:

$$\sigma = \sigma_p \sum_i (1 + m_i [VF_i])$$

where: $[VF_i]$ is the volume fraction of aggregate grade i and:

$$m_i = 3\alpha_i \frac{2(\alpha_i - 1)\sigma_{ITZ} - (2\alpha_i + 1)\sigma_p}{2(\alpha_i - 1)\sigma_{ITZ} + 2(2\alpha_i + 1)\sigma_p} \quad \text{where} \quad \alpha_i = \left[1 + \frac{\text{ITZ thickness}}{\text{average radius of grade } i}\right]^3$$

in which: σ_{ITZ} is the conductivity of the ITZ. A thorough interpretation of m_i is given in Garboczi, et al., [30]; in short, m_i becomes less negative (and may become positive) as the relative conductivity and thickness of the ITZ increases. An alternate formulation of the composite conductivity is based on Bruggeman's EMT results [29], with percolation effects by Mclachlan, et al., [36]:

$$\sigma = \sigma_p \prod_i \left(1 - \frac{[VF_i]}{[VF_C]}\right)^{-m_i [VF_C]}$$

where: $[VF_C]$ is the critical volume fraction corresponding to percolation. For this example analysis, a simple model for the ITZ porosity is used, where the ratio of ITZ and paste porosities is specified by a constant, β.

An inverse relation is obtained using the aforementioned model with the experimental data as follows. In Figure 1, the independent variable was the aggregate volume fraction, [VF], which is directly related to the experimental design parameters; the dependent variable is the conductivity of the concrete, σ, which is directly related to the data measurements. The problem with this choice of variables is that there is no one-to-one functional relation between conductivity and aggregate volume fraction, as clearly indicated in the figure. In the inverse analysis, the conductivity of the paste, σ_p, is chosen for the dependent variable and the w/c ratio for the paste is chosen for the independent variable. A one-to-one functional relation between these variables is expected because the relation is an intrinsic property of the paste, independent of the aggregate. However, since neither the conductivity of the paste nor the w/c ratio for the paste can be directly or indirectly measured, they must be estimated by inverting the preceding analytical models. In the models, the experimental data is substituted for the conductivity of the concrete, σ, and the parameters for the ITZ, including its thickness, relative conductivity, σ_{ITZ}/σ, and relative porosity, β, are obtained by statistical means. Two results are shown in Figure 2. In Figure 2(a), the relation between paste w/c ratio and conductivity is plotted without including ITZ in the analytical model, and in Figure 2(b), the same relation is plotted with ITZ thickness of 0.02 mm, relative conductivity $\sigma_{ITZ}/\sigma = 6$, and relative porosity, $\beta = 1.7$.

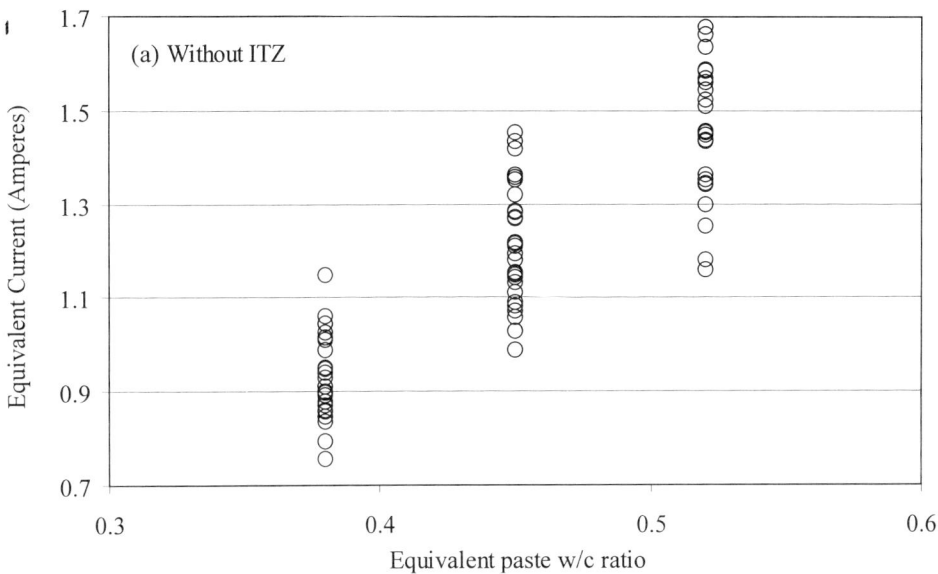

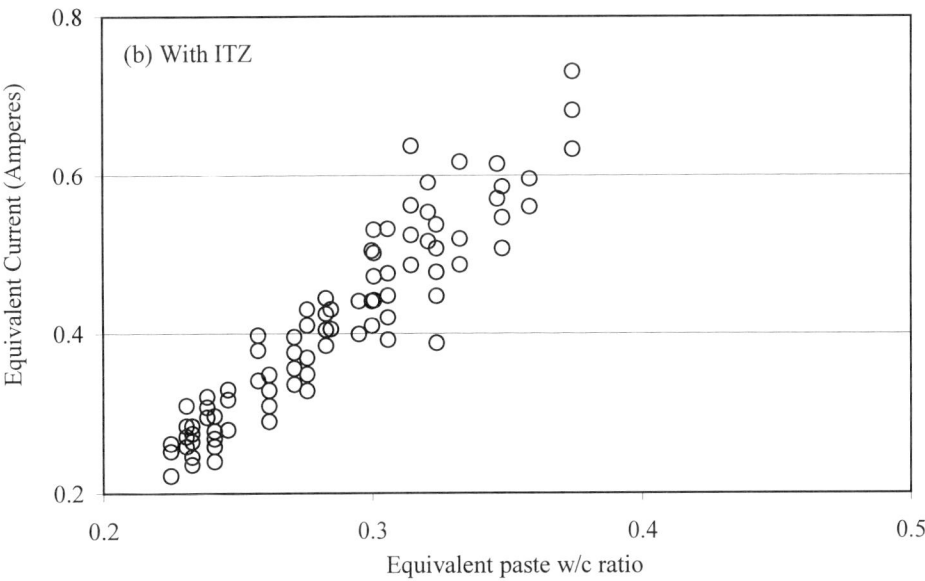

Figure 2 (a) and (b) showing inferred relations between cement paste w/c ratio and conductivity

21. MONTEIRO, P J M, GJØRV, O E AND MEHTA, P K. Microstructure of the Steel-Cement Paste Interface in the Presence of Chloride, Cement and Concrete Research, Vol. 15, 1985, pp. 781-784.

22. MAJI A AND SHAH S P. Process Zone and Acoustic Emission Measurement in Concrete, Experimental Mechanics, 1988, pp. 27-33.

23. LANDIS, E N, SHAH, S P AND LI, Z. Application of Quantitative NDE to Basic Fracture Research of Concrete, Review of Progress in QNDE, Vol. 13, 1994, pp. 2123-2130.

24. OHTSU, M, OKAMONOT, T AND YUYAMA, S. Moment Tensor Analysis of Acoustic Emission for Cracking Mechanisms in Concrete, ACI Structural Journal, 1998, pp. 87-95.

25. HEARN, S W AND SHIELD C K. Acoustic Emission Monitoring as a Nondestructive Testing Technique in Reinforced Concrete, ACI Structural Journal, 1997, pp. 510-519.

26. SHAH, S P, SWARTZ, S E AND OUYANG C. Fracture Mechanics of Concrete: Application of Fracture Mechanics to Concrete, Rock and Other Quasi-Brittle Materials, John Wiley & Sons, Inc., New York, 1995.

27. SAHIMI, M. Flow and Transport in Porous Media and Fractured Rock. VCH, 1995, pp. 1-10 and 51-53.

28. MAXWELL, J C. Electricity and Magnetism, 3rd edition, Oxford, 1892, pg. 440.

29. BRUGGEMAN, D A, Annals of Physics, Vol. 24, 1935, pg. 636.

30. GARBOCZI, E J, SCHWARTZ, L M AND BENTZ, D P. Modeling the Influence of the Interfacial Zone on the DC Electrical Conductivity of Mortar. ACBM Journal, Elsevier Science, 1995.

31. RAYLEIGH, LORD, On the Influence of Obstacles Arranged in Rectangular Order upon the Properties of a Medium. Philosophical Magazine, Vol. 34, 1892, pp. 481-502.

32. HANSHIN, Z. Analysis of Composite Materials, Journal of Applied Mechanics, Vol. 50, 1983, pp. 481-505.

33. TORQUATO, S AND RINTOUL, M D. Effect of the Interface on the Properties of Composite Media. Physics Review Letters, Vol. 75, 1995, pp. 4067-4070.

34. CHENG, H AND TORQUATO, S. Effective Conductivity of Periodic Arrays of Spheres with Interfacial Resistance. Proceedings of the Royal Society of London A, Vol. 453, 1997, pp. 145-161.

35. GARBOCZI, E J AND BENTZ, D P. Analytical Formulas for Interfacial Transition Zone Properties. Materials Research Society, Fall 1996, preprint.

36. MCLACHLAN, D S, BLASZKIEWICZ, M AND NEWNHAM, R E. Electrical Resistivity of Composites. Journal of American Ceramic Society, Vol. 73, No. 8, 1990, pp. 2187-2203.

37. CHRISTENSEN, B J, COVERDALE, R T, OLSON, R A, FORD, S J, GARBOCZI, E J, JENNINGS, H M AND MASON, T O. Impedance Spectroscopy of Hydrating Cement-Based Materials: Measurement, Interpretation, and Application. Journal of American Ceramic Society, Vol. 77, No. 11, 1994, pp. 2789-2804.

38. OLLIVIER, J P AND MASSAT, M. The Effect of the Transition Zone on Transfer Properties of Concrete. Interface Transition Zone in Concrete, Ed. J C Maso, RILEM Report 11.

39. WINSLOW, D F AND COHEN M D. Percolation and Pore Structure in Mortars and Concrete. Cement and Concrete Research, Vol. 24, 1994, pp. 25-37.

40. JAISWAL, S, IGUSA, T, STYER, P, KARR, A F, AND SHAH, S P. Influence of Microstructure and Fracture on the Transport Properties in Cement-Based Materials. Brittle Matrix Composites, International Conference on Brittle Matrix Composites, Eds. Brandt, A. M., et al., Warsaw, Poland, October 6-8, 1997, pp. 199-220.

THEME ONE: BASIC MECHANISMS

Keynote Paper

WHY ARE AGGREGATE/MATRIX INTERFACES THAT THICK?

J M J M Bijen
INTRON BV
M de Rooij
Delft University of Technology
Netherlands

ABSTRACT. The interfacial zone between the aggregate interface and the bulk of the cement matrix is remarkably thick. Modelling has shown that the size of this region cannot be explained by particle packing at the 'wall' alone. By observing through a microscope a sealed droplet of a cement suspension it has been proved that the dispersed cement particles in the suspension do contract leaving a water layer at the outskirts of the droplet. This phenomenon is known from colloid chemistry as syneresis. Flocculated systems improve their free energy by a contraction process in which more interparticle contacts are realised. During contraction water is expelled from the flocculated system. In the case of well-dispersed, non-flocculated cement suspensions using superplasticizers the contraction is substantially less. These results appear to correspond to the size of the interfacial zone observed in hardened concrete. It is believed that this phenomenon can explain the size of the transition zone. Experiments have shown that the decrease in the thickness of the transition zone observed where fine pozzolans have been added cannot be explained by this phenomenon.

Keywords: Cement, Suspension, Flocculation, Syneresis, Coagulation, Admixtures, Additions, Interface, Transition, Particle-packing.

Professor Jan M J M Bijen is director of INTRON BV the Dutch institute for quality assessment in the building industry. He is also part-time professor in materials science at the Faculty of Civil Engineering and Geosciences of Delft University of Technology. He specialises in the durability and sustainability building materials.

Mario de Rooij is a PhD student at the section of Civil Engineering Materials of the Faculty of Civil Engineering and Geosciences of Delft University of Technology in the Netherlands. He is investigating the contraction of cement suspensions in relation to the transition zone between aggregate and cement matrix in concrete.

INTRODUCTION

The composition of the interfacial zone in the cement matrix at the aggregate surface differs substantially of that of the bulk of the cement matrix in concrete. Generally, it is weaker and more porous. It is believed that this region greatly influences the bulk properties of concrete, such as strength, transport ability for water and ions in the pore water, resistance to deteriorating mechanisms such as alkali-silica reaction and sulphate attack. When measured, the thickness of the interfacial region is often about 40-50 µm, while the mean spacing between aggregates for a well-proportioned concrete is about 150-100 µm. Thus a large fraction of the cement matrix belongs to the interfacial zone.

The literature gives a number of reasons why the interfacial zone is that thick. These are:

- Bleeding, e.g. segregation due to gravitational forces
- A particle packing wall effect, solid particles can only touch the aggregate surface with their edges.
- Adsorption of a water layer both at the surface of the aggregate and at te cement particle.
- Effect of drying shrinkage during preparation of samples for analysis
- Flocculation of cement particles

This paper concentrates on the particle packing wall effect, flocculation, and a new mechanism investigated at Delft University of Technology: syneresis. Also considered are the effects of fine fillers and pozzolanic additions on the interfacial zone.

THE WALL EFFECT

The particle packing wall effect is often mentioned as one of the reasons for the existence of a porous interfacial zone. Modelling of this effect such as with the Particle program [1] does show that there will be a gradient in solid density from the aggregate surface into the cement matrix. Because for Portland clinker particles hydration products grow concentrically inwards and outwards from the contours of the original particles this gradient will also remain after hydration [2]. Calculations with Particles, see Figures 1 and 2, do show however that for a typical cement grain size distribution this only accounts for an interfacial zone thickness of about 10 µm.

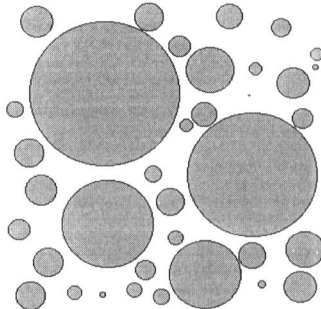

Figure 1 Particles packing restricted by walls

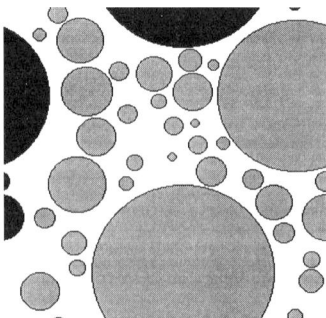

Figure 2 Particles packing not restricted by walls

FLOCCULATION

Normal, neat cement suspensions are flocculated. Figure 3 shows a coagulated and a well-dispersed system. Well-dispersed systems can be achieved by applying detergents, which change the zeta potential of the cement particles. Normal cement suspensions have a high ionic strength.

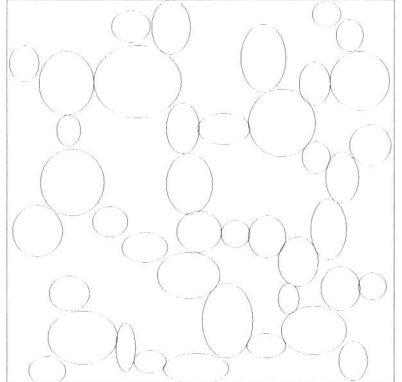

Figure 3 a)   Flocculated cement particles   b) Well-dispersed cement particles

Applying the DVLO theory, Yang et al [3] have recently show that flocculation will occur between zeta potentials of –20 mV and 20 mV. The zeta-potential can effectively be changed to a level outside this region by using detergents as superplasticizers and air entraining agents.This phenomenon does explain the effects of admixtures on the sedimentation of cement pastes and concrete (bleeding) as shown by Neubauer et al [4]. In well-dispersed systems the heaviest particles do settle first. In flocculated systems the flocs settle with the smaller particles entrapped in them, so that no particle size gradient is formed.

DEFINITION OF SYNERESIS

An investigation by the authors into the reasons for the existence of the interfacial zone has revealed that another phenomenon known from colloid chemistry can explain the existence of the water-rich layer: syneresis. When a system undergoes rapid flocculation the result is a loose floc in which most particles tend to be linked to two or three other particles. The structure of these flocs is tenuous and contains a substantial amount of entrapped water. The free energy can be decreased through the formation of a greater number of contacts. This entails a contraction of the dispersed phase. The volume of the dispersed phase is decreased and water is spontaneously expelled from the flocs. This phenomenon was first referred to as syneresis by Thomas Graham [5]. Figure 4 shows schematically the syneresis process for a cement suspension.

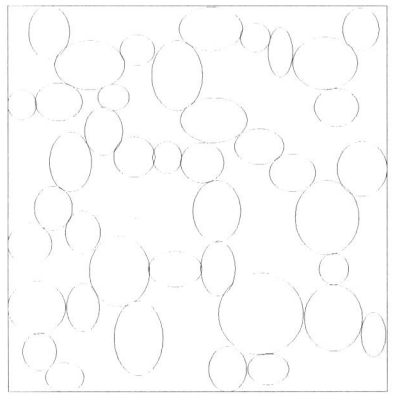 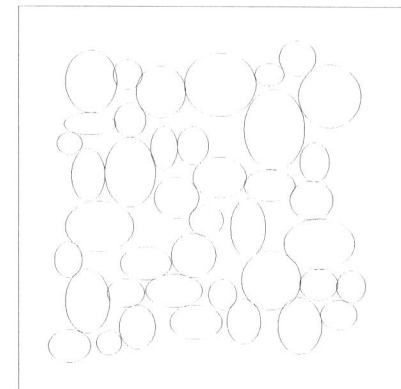

Figure 4 a) Flocculated cement particles   b) Contacted flocculated cement particles
                                                (syneresis)

An example of syneresis is the formation of a blood clot. When blood flows out of a vessel in a cut and it is exposed to air, a complex system of reactions turns albumins into a network of fibrin fibres. Blood cells become entrapped in this network.

Meanwhile the blood serum is expelled. Another example can be found in the cheese-making process. Gels formed from milk by renetting or acidification under quiescent conditions show syneresis. The gel contracts, while expelling liquid (whey). Syneresis is also suggested for systems similar to cement suspensions such as for silicate-alumina grout [6].

It is important to realise that both flocculation and syneresis occur within a few hours after mixing.

SYNERESIS OF CEMENT PASTES

The authors were able to demonstrate the existence of contraction of the dispersed cement particles (also known as syneresis) in cement suspensions and pastes using a relatively simple technique. A droplet of cement suspension with a water/cement ratio usual for concrete is placed on a slide object holder.

To prevent evaporation and carbonation the cement drop is sealed off from the atmosphere using double-sided tape and a cover sheet.

The bottom plate is of PMMA instead of glass to prevent any interaction with the carrier. Glass has a high surface tension and will be wetted by the water in the suspension, thus interfering with the contraction process. The droplet is studied under optical microscope.

The experimental set up is shown in Figure 5.

Figure 5 Schematic depiction of the sealing of the cement suspension droplet for observation under the microscope

The contraction of the cement particles is measured using image analysis. Figure 6 shows a photograph of a part of the droplet directly after mixing and 1 hour respectively.

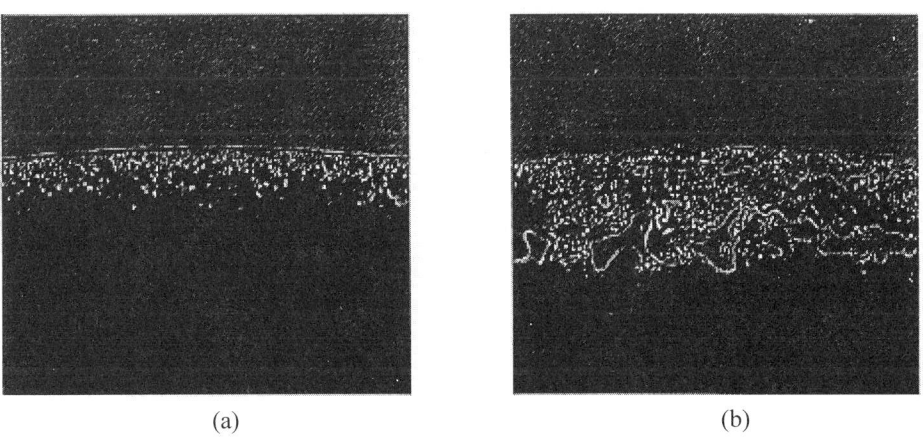

Figure 6 Photograph of cement suspension droplet (a) directly after mixing and (b) after 60 minutes

In Figure 7 the relative contraction of the dispersed cement phase is shown for a CEM I 32,5R cement paste with a water/cement ratio of 0.5 with and without 1.5% m/m cement superplasticizer (Tillmann OFT3) on cement.

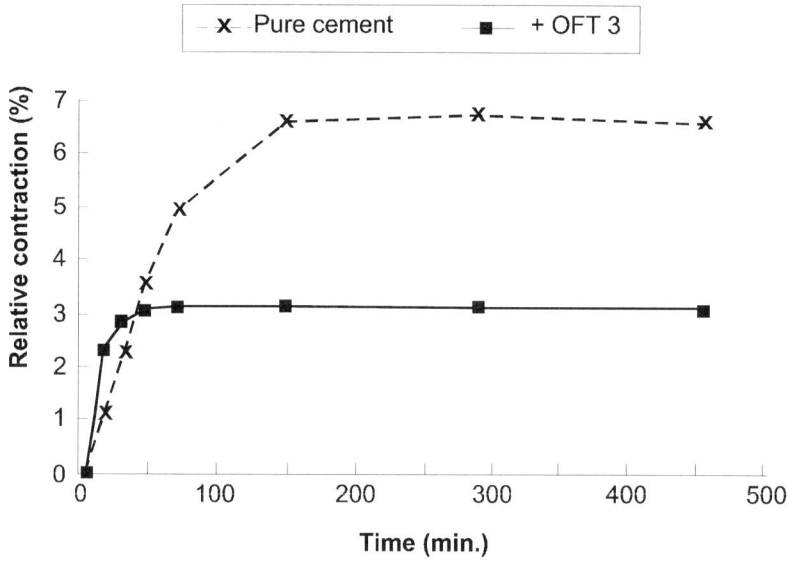

Figure 7   Relative contraction of cement particles in pastes

In the latter case the retreat of the cement particles is substantially less. Kim and Robertson did show [7] that polyvinyl alcohol (PVA) gave a well-dispersed, non-flocculated system with a substantial reduction of the thickness of the interfacial zone.

According to these investigators this is due to a wall effect arising from the flocculation of cement particles. The authors are of the opinion, this wall effect is due flocculation followed by syneresis. Our observations also correspond with those of Zimbelmann [8], who found a water-rich layer at the aggregate surface.

EFFECTS OF ADDITIONS

In a previous investigation of our section the effect of additions on the interfacial zone was studied [10].

In agreement with findings of others it was found that additions such as silica fume and fly ash do give a substantially less thick interfacial zone, see for instance Figures 8 and 9 for fly ash.

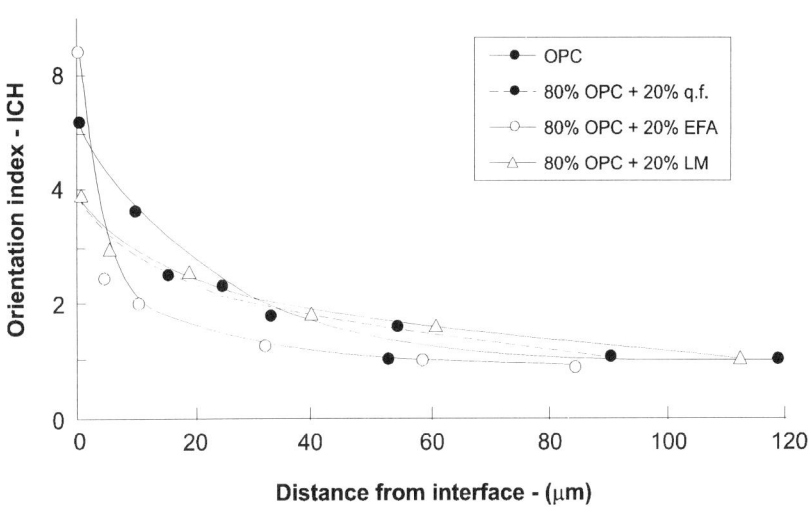

Figure 8 Degree of orientation of CH at the paste –
"coarse aggregate" interfacial zone (7 days old) [9]

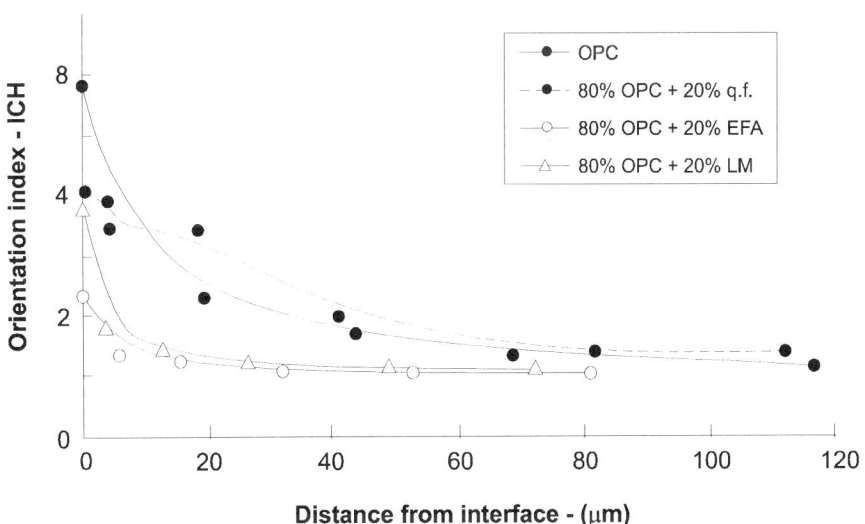

Figure 9 Degree of orientation of CH at the paste-
"coarse aggregate" interfacial zone (28 days old) [9]

In the present investigation the microscope technique discussed above was used to investigate the effect of fly ash and silica fume on syneresis. We did not observe a substantial effect on the contraction of the dispersed cement phase. This suggests that these additions do not have a substantial effect on syneresis.

Studying the effects of fly ash in more detail, it is obvious from our own results, e.g. figure 8, as well as results from others that a slow densification of the interfacial zone occurs over time and not a fast one that could be due to syneresis. The densification, which occurs over time, may be related to the reaction mechanism of pozzolanic – cementitious systems as discussed by Bakker [10]. His hypothesis is that reaction products are formed between cement and pozzolan particles instead of on and in the cement particle as in neat Portland cement pastes. Of course, the presence of fine particles will also influence the contribution of the particle packing wall effect to the size of the interfacial zone as discussed in a previous paragraph.

CONCLUSIONS

1. The results of an investigation indicate that the substantial thickness of the interfacial zone in the matrix at the aggregate interface may be explaned by syneresis of the cement paste.

2. Pastes with detergents, which change the zeta-potential substantially, exhibit less syneresis.

3. Additions like fly ash and silica fume do not show major changes in the extent of the contraction of the dispersed cement particles phase.

4. The observed decrease of the thickness of this zone in time by these additions is likely to be caused by the precipitation of reaction products remoted from the original cement particle.

ACKNOWLEDGEMENTS

The authors would like to express their gratitude to professor G. Frens of the Faculty of Chemical Engineering and Material Science of Delft University of Technology for drawing their attention to the phenomenon of syneresis.

REFERENCES

1. SCHLANGEN, E., Calculations with Particles, 1997 (Particles is a proprietary name of FEMMASSE BV, The Netherlands).

2. VAN BREUGEL, K., Simulation of Hydration and Formation in hardening Cement-based Materials, dissertation, Delft University of Technology 1991, ISBN 90-9004618-6.

3. YANG, M., NEUBAUER, C.M., JENNINGS, H.M., Interparticle Potential and Sedimentation Behaviour of Cement Suspensions, Advance Cement base Materials 5, 1997, pp 1-7.

4. NEUBAUER, M. C., YANG M., JENNINGS, H.M., Interparticle Potential and Sedimentation Behavior of Cement Suspensions: Effects of Admixtures, Advanced Cement Based Materials 8, 1998, pp 17-27.

5. KRUYT, H. R., Colloid Science II, Elsevier Publishing Company Inc., New York, 1949, p 573.

6. JEFFERIS, S. A., SHEIKH BAHAI, A., Geotechnique 45, 1995 pp 131-140.

7. KIM, J-H., ROBERTSON, R. E., Effects of Polyvinyl Alcohol on Aggregate-Paste Bond Strength and the Interfacial transition Zone, Advanced Cement Based Materials 8, 1998, pp 66-76t.

8. ZIMBELMANN, R., A Contribution to the Problem of Cement-Aggregate Bond, Cement and Concrete Research 15, no 5, 1985, pp 801-808.

9. LARBI, J.A., The Cement Paste-Aggregate Interfacial Zone in Concrete, dissertation, Delft University of Technology 1991.

10. BAKKER, R.F.M., Permeability of Blended Cement Concretes, ACI SP 79, 1983, paper SP 79-30, pp 589-605.

CORROSION AND BOND CHARACTERISTICS OF REINFORCING STEEL IN NORMAL PORTLAND AND FLY ASH CONCRETES

L Amleh
M S Mirza
McGill University
Canada

ABSTRACT. This paper presents the results of a preliminary study involving 14 tension specimens, subjected to seven different levels of accelerated corrosion (zero corrosion to extensive corrosion involving 9 mm wide longitudinal cracks and about 17.5 percent loss of cross-sectional area). The response was evaluated in terms of tension stiffening, cracking behaviour, chloride ion profile, influence of cracking due to corrosion, load-deformation and bond characteristics. The relative bond efficiency at the steel-concrete interface was determined based on the crack spacing. A 4 percent weight loss caused a decrease of 9 percent in bond strength, however, a 17.5 percent weight loss led to a drastic 92 percent loss in bond strength.

In addition, partial results of a detailed investigation involving 192 lollipop specimens made from six different concrete mixes (Portland and high alumina cements and pulverized-fuel ash), with each specimen reinforced with a single #20 bar. Four different concrete covers (25, 50, 75 and 100 mm) were used. A total of 144 specimens were subjected to accelerated corrosion (5% NaCl solution and 5 volts imposed voltage) for increasing periods of time to attain four stages of corrosion ranging from no corrosion to extensive corrosion at the steel-concrete interface. Besides visual examination, a series of electrochemical and other tests were undertaken to study the various aspects of steel bar corrosion.

Keywords: Accelerated corrosion, Bond deterioration, Chloride ion profile, Corrosion level, Electrochemical testing, Pulverized-fuel ash (PFA), Lollipop specimens, Longitudinal and transverse cracks, Reinforced concrete.

Ms Lamya Amleh is a PhD candidate at McGill University. Her interests are in the area of deterioration and renovation of infrastructure, with emphasis on corrosion of reinforcing steel embedded in concrete. Her PhD program deals with deterioration of bond with increasing levels of corrosion on both laboratory specimens and in structural members in the field.

Professor M Saeed Mirza is a Professor in the Department of Civil Engineering and Applied Mechanics at MGill University, Montreal, Canada. His research focuses on behaviour and design of concrete structures, and on durability of concrete infrastructure. Professor Mirza has been an author and coauthor of several publications, has been a member of many related committees, and has been frequently invited to be an expert witness.

INTRODUCTION

Corrosion of reinforcing steel in concrete infrastructure has become a very serious problem around the world, requiring billions of dollars for renovation and replacement. The alkaline concrete environment normally provides protection against corrosion of reinforcing steel through a protective layer on the surface of the steel bar, and it can get damaged or removed by ingress of chlorides or carbon dioxide. Availability of moisture and oxygen near the bar surface subsequently results in the corrosion of the steel bar. Andrade et al [1], and Amleh and Mirza [2] observed that a 5% loss of cross-sectional area, even with cracking and minor splitting did not influence the structural performance significantly. This situation represents an early stage of corrosion deterioration, and it can be improved by appropriate repairs. They also noted that a 10-25% loss of steel cross-sectional area in critical locations of the structure can deplete the service life because of loss, or reduction of structural integrity. This situation can become significantly dangerous due to the loss of bond at the steel-concrete interface.

It is generally accepted that most of the bond resistance arises from the mechanical interlocking of the steel bar to the surrounding concrete and the mode of bond failure depends on the embedment length, concrete cover thickness and the strength of the concrete around it. The thickness and impermeability of the concrete cover over the steel reinforcement provide both physical and chemical protection by providing an alkaline and electrically resistive barrier against the ingress of moisture, oxygen, carbon dioxide, chlorides and other aggressive elements. This implies that adequate protection against steel corrosion can be obtained by using a well compacted and cured concrete mix with dense aggregates and a low water-cement ratio. This protection can be improved further using supplementary cementing materials in the concrete mixture.

Incorporation of pulverized-fuel ash (PFA) in concrete influences the permeability of the concrete cover. Properly proportioned, compacted and cured PFA concrete is generally less permeable at later ages than the corresponding plain concrete. However, the higher time period for pozzolanic reactions leads to a higher level of permeability in PFA concrete at early ages which enhances the risk of steel corrosion in any premature exposure of the concrete to aggressive agents.

RESEARCH OBJECTIVES

This research program is aimed at studying the influence of corrosion on bond characteristics between the reinforcing steel and the surrounding concrete. After a preliminary series of 14 tension tests at seven different corrosion levels to establish the influence of accelerated corrosion on bond response, a detailed investigation is currently in progress to study the influence of four levels of corrosion on bond characteristics, using lollipop specimens and accelerated corrosion techniques involving six different concrete mixtures made from Portland cement (PC) with and without PFA, and four different concrete cover thicknesses (25, 50, 75 and 100 mm).

It should be noted that inititation and propagation of corrosion in steel reinforcing bars embedded in the concrete can occur in a few to several years, depending on the structural details and the aggressiveness of the microclimate. This period can be telescoped in the laboratory to develop useful data in a relatively short period by using accelerated corrosion techniques, such as the ones developed at McGill University. However, correspondence needs to be established between the laboratory and field periods of corrosion, and between the responses observed in the laboratory tests and the behaviour of reinforcing bars in deteriorated field structures. Detailed investigations are planned on the abandoned Dickson Bridge in Montreal and the Perly Bridge in Hawkesbury to generate this data. These results will be reported later.

PRELIMINARY STUDY OF BOND DETERIORATION WITH CORROSION LEVEL

Amleh [3] undertook a preliminary series of tests on 14 tension specimens, 100 mm diameter by one meter long, w/c ratio = 0.45, 28-day compressive strength of 25 MPa, and reinforced with a single #20 steel bar, where #20 bar denotes a 19.5 mm diameter reinforcing bar, (f_y = 400 MPa) to study the bond characteristics at the steel-concrete interface for seven different selected corrosion levels, based on the loss of bar cross-sectional area and width of the longitudinal cracks. These ranged from no corrosion to extensive corrosion of the reinforcing bars, with significantly large longitudinal cracks at higher levels of corrosion. In each case, the specimen response was evaluated in terms of tension stiffening, cracking behaviour, influence of pre-existing cracks due to corrosion, load-deformation and bond stress characteristics. The relative bond efficiency at the steel-concrete interface was determined based on the experimental crack spacing, along with the evaluation of the chloride ion profile within the concrete cover over the bar.

The tension specimens were cast in specially designed plastic moulds, with sockets at both ends to hold the protruding bar (150 mm on each side) in a correct position and to provide a uniform concrete cover (40 mm). Along with the auxiliary specimens, the tension specimens were wet-cured for 28 days after removal from the moulds after 48 hours. They were then immersed in an electrolyte tank (5% NaCl solution by weight of water which was changed weekly) and subjected to an imposed voltage of 5 volts. The reinforcing bar (#20) acted as the anode while a bare bar placed in the tank acted as the cathode. The current and voltage were measured at 48 hour intervals. Once the desired level of corrosion was attained, it was tested in tension in a servo-controlled hydraulic universal MTS machine (1000 kN capacity) using displacement control. The elongation of the steel bar was measured using a built-in LVDT and a specially designed extensometer.

Reinforcing Steel Corrosion Levels

The seven corrosion levels were established based on the onset, propagation and widening of the longitudinal cracks with the average widths ranging from 0.1 mm to 9 mm, representing the first and the last stages of corrosion, respectively. The classifications were further affirmed by the measurement of the bar weight loss after the test and the loss in cross-sectional areas based on the experimental ultimate strength and the assumption that the yield strength is constant in both the corroded and the uncorroded bars.

Chloride Ion Profile

The chloride ion content of the powder samples obtained by drilling to depths of 15 mm, 30 mm and 40 mm from the specimen surface, was determined using the Volhard Method of the

British Standard 1881 Part 124: 1998. The uncracked regions displayed normal variations from a maximum value near the surface to a minimum at the bar surface. However, the profile patterns at the cracks were significantly different due to the unimpeded ingress of chlorides through the cracks. This also increased the metal loss and the intensity of pitting, which was similar to that observed by Rasheeduzzafar[4].

Influence of Corrosion on Bond

As the steel bar corroded, the increased volume of the corrosion products resulted in "bursting" pressure which caused longitudinal splitting cracks in the specimens, with the crack width increasing with the corrosion level. This resulted in the breakdown of cohesion, adhesion and friction at the steel-concrete interface, excepting at low corrosion levels, where there was no longitudinal cracking, and the corrosion products had a beneficial effect of improving the bond characteristics at the steel-concrete interface [5]. At higher corrosion levels, the steel bars displayed localized pitting and loss of some of the ribs over the bar length, thereby weakening the rib-concrete mechanical interlocking force transfer mechanism. An examination of the number and spacing of the transverse cracks showed that as the level of corrosion increased, the transverse crack spacing also increased, reflecting the deterioration of bond characteristics at the steel-concrete interface [2].

Load-Elongation Response

The load-elongation curves clearly reflected the effect of tension cracking which in turn depended on the quality of bond. As the level of corrosion increased, the load deformation curves shifted close to the bare bar response, indicating a deterioration of bond (Figure1). The fifth and sixth levels of corrosion, with bar weight losses of 12 and 17.5 percent, respectively, showed hardly any contribution of concrete in terms of tension stiffening, indicating a large deterioration in bond.

Tests on these corroded bars also displayed considerably reduced ductility. The experimental ultimate steel strength was translated into a reduced cross-sectional area, assuming that the yield strength did not change with corrosion. This showed cross-sectional area losses ranging from 1.7 percent for the first level of corrosion to about 20 percent for the last corrosion level.

The results of the relative bond strength were evaluated assuming the average bond stress to be constant between the cracks and the standard equation ($u=(df_y)/4\ell$, where u is the average bond stress, d is the bar diameter, f_y is the steel yield strength and ℓ is the length of the bar). The results showed that a 4 percent weight loss caused a 9 percent decrease in bond strength, however, a 17.5 percent weight loss led to a drastic 92 percent loss in bond strength, which is due to the loss of the mechanical interlocking at the deteriorated ribs (Figure 2).

CURRENT STUDY

Pulverized-Fuel Ash Concretes

Over the past 15 years, Malhotra [6] has undertaken detailed research on high volume pulverized-fuel ash (HVFA) concretes at the CANMET Laboratories, Ottawa. He noted that

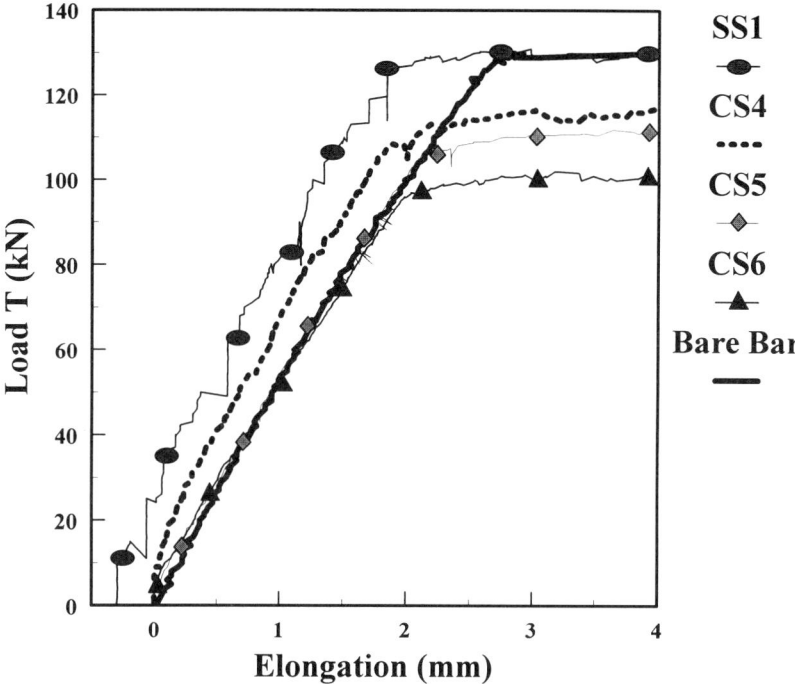

Figure 1 Tensile force-elongation responses for Specimens SS1, CS4, CS5, CS6

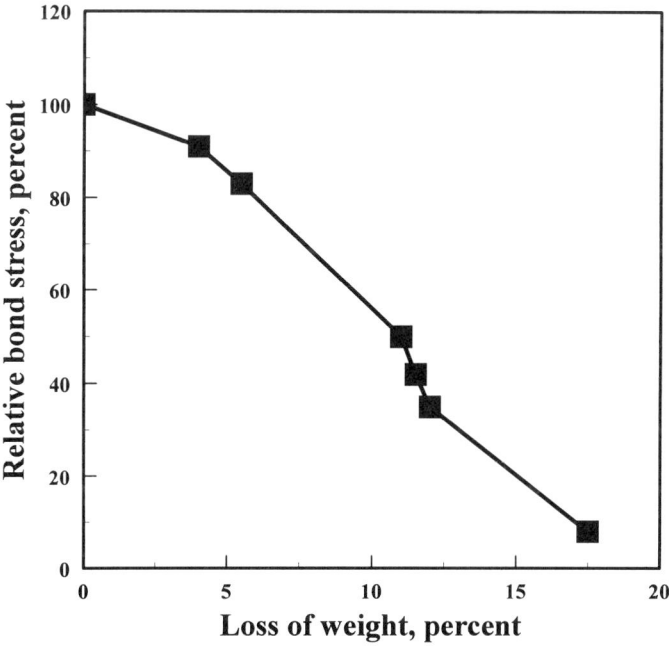

Figure 2 Relative bond stress versus reinforcing steel bar loss of weight

the HVFA concretes demonstrated excellent characteristics for resistance against freezing-thawing cycles, penetration to chloride ions, deicing salts, carbonation and alkali-aggregate reactivity. The excellent strength, permeability and durability characteristics of HVFA concrete can be beneficial in bridge decks and parking structures, which have shown considerable distress and deterioration in several cold climate countries due to the application of deicing salts in the winter. The same is true for structures in seawater environment. Despite the considerable research on HVFA concretes, their influence in inhibiting or impeding corrosion of the steel reinforcement and its effects on bond characteristics at the steel-concrete interface have not received adequate attention and need to be investigated.

Laboratory Program

Amleh [7] is presently involved in a detailed research program aimed at testing 192 lollipop specimens (300 mm long) with six different concrete mixes (two PC, three PFA and one high alumina cement (HAC) concretes) and four different concrete cover thicknesses (25, 50, 75 and 100 mm), with each specimen reinforced with one symmetrically placed #20 bar, protruding at one end only. The PFA used are from Point Tupper PFA1, Thunder Bay PFA2 and Sundance PFA3 with the PFA-cement ratio being 0.58, and a water-cementitious materials ratio of 0.32 (see Table 1 for details). The two PC mixes had water-cement ratios of 0.32 and 0.42, while for the HAC concrete, the water-cement ratio was 0.37.

Besides the uncorroded control specimens, the specimens are subjected to accelerated corrosion by immersion in a 5 percent solution of sodium chloride and an applied voltage of 5 volts. The specimens are kept immersed until the desired level of corrosion is attained, corresponding to four stages of corrosion, ranging from no corrosion to a very high level of corrosion.

Table 1 Concrete mixes

MIX.	CONCRETE TYPE	W/C+SC M	QUANTITIES (kg/m^3)					A.E.A.*** (mL/m^3)	STRENGTH, MPa
			Water*	Cement	PFA	Agg.	SP**		
C1	PC/PFA1	0.32	118	156	216	1858	5.5	422	52.2
C2	PC/PFA2	0.32	117	153	213	1815	2.5	228	44.1
C3	PC/PFA3	0.32	121	157	218	1803	6.0	558	59.2
C4	PC1	0.32	119	371	0	1877	10.1	401	51
C5	PC2	0.42	154	168	0	1244	1.3	149	49.5
C6	HAC	0.37	165	451	0	1823	0.0	0	49.7

PC- Portland cement   PFA- Pulverized-fuel ash   HAC- High aluminium cement
*   including water in the superplasticizer **   superplasticizer, naphthalene based
***   air-entraining admixture

Electrochemical tests (half-cell potential, Tafel, linear polarization, and potentiodynamic tests) have been used to decide whether a specified stage of corrosion deterioration has been attained, and to study the various corrosion characteristics.

Details of the various tests, test data, analysis methods and results are presented by Amleh [7]. Some typical results are presented here.

Corrosion Current

The measured current due to an applied voltage is an indicator of corrosion activity in the specimens. A typical current vs immersion time data set for all six concretes and a cover thickness of 50 mm are shown in Figure 3. This and the other data, not reported here, show that initially the measured current is small, however, it increased with the exposure time for the PC concrete mixture with a w/c ratio of 0.42. This current is a function of the concrete cover thickness and the resistivity of the concrete which decreases with its deterioration.

Half-Cell Potential Tests

The results of the half-cell potential measured using a copper-copper sulphate ($Cu-CuSO_4$) half cell are shown in Figure 4 for specimens with a 50mm thick concrete cover for all six concrete mixes. This sample and other data show a lower probability of corrosion for PFA concretes (C1- PFA1, C2- PFA2, C3- PFA3) and PC concrete (C4, w/c ratio = 0.32). This is also confirmed from the corrosion initiation time for the different concretes and cover thicknesses. All of the data (not reported here) clearly show the superior corrosion protection of the PC concrete with a w/c ratio of 0.32, compared with that for the concrete with a higher w/c ratio (0.42, C5).

Corrosion Rate

Typical variations of the corrosion rate with time (in mm per year) for the specimens made with the six concretes and 50mm thick concrete cover are shown in Figure 5. These and other results show that the three PFA and the PC (w/c ratio = 0.32) concretes provide good protection against chloride ion penetration which is reflected in the lower corrosion rates. Similarly partial replacement (58%) of cement with PFA has a significantly beneficial effect on the corrosion initiation time, clearly showing the superior passivating characteristics of the PFA and PC concretes (w/c ratio=0.32) over those for the PC concrete (w/c ratio = 0.42) and the HAC (C6) concrete.

Pullout Force - Slip Responses

Typical load-slip responses from the pullout tests on three Point Tup[er fly ash mixture are shown in Figure 6 for specimens with 50 mm thick concrete covers for Stages 1 (no corrosion) and 2 (precracking stage). It can be noted that the bond strength increases with corrosion up to 1 percent, however, with further increase in corrosion bond strength decreases significantly. This and other data for tests undertaken on specimens with 25 and 50 mm cover in Stages 2,3 and 4 so far (and not reported here) show a linear relationship between the applied pullout force and the slip up to about 75% of the ultimate load when the curve becomes nonlinear due to deterioration of bond. Similarly, the steel reinforcement in the PC concrete specimens with 50, 75 and 100 mm covers failed due to yielding of the steel reinforcement for Stages 1 and 2. Thus, there was no breakdown of bond during these early stages of corrosion for specimens with larger concrete covers. Some results from these tests with 25 and 50 mm concrete covers are summarized in Table 2.

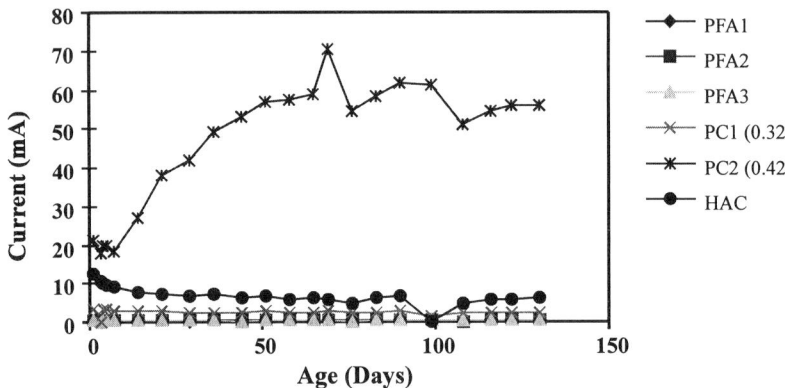

Figure 3 Current readings for specimens with 50mm concrete cover

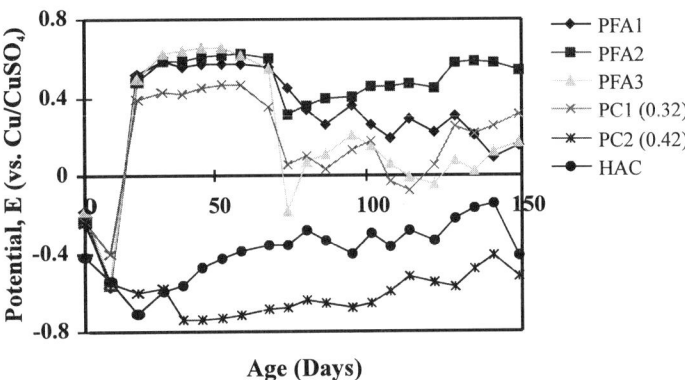

Figure 4 Half-cell potential data for typical lollipop specimens with 50mm concrete cover

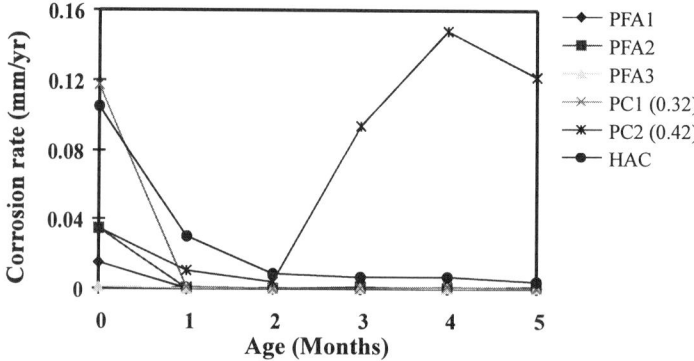

Figure 5 Corrosion rate vs time for 50mm cover depth specimens

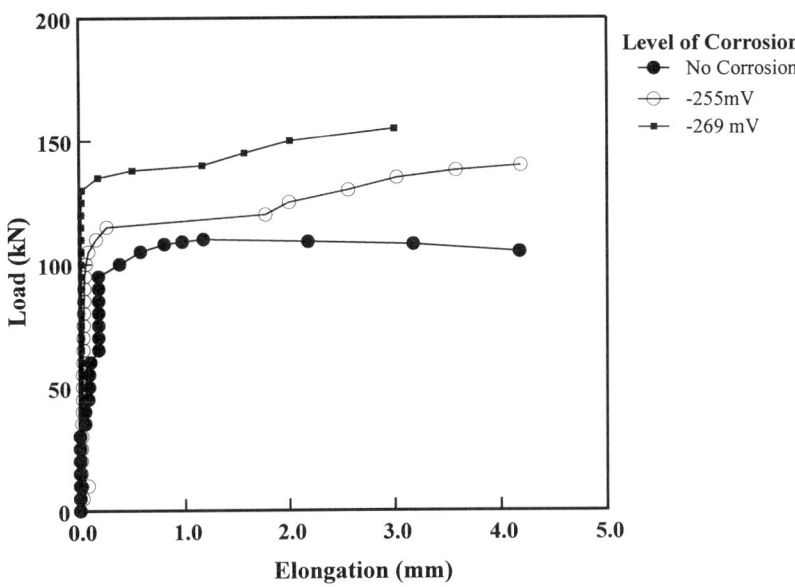

Figure 6 Load versus elongation for (PFA1) concrete mix, 50mm cover

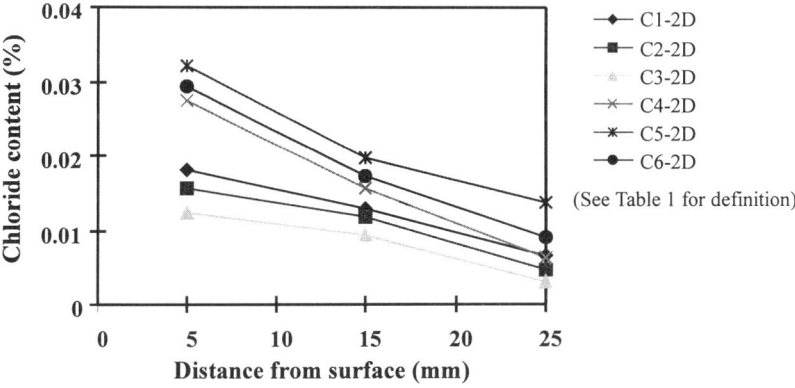

Figure 7 Chloride penetration for 25mm cover depth sections

Table 2 Bond performance for all the concrete mix, 25 and 50 mm concrete cover

CONCRETE MIXTURE	SPECIMEN NUMBER	FAILURE LOAD (KN)	HALF-CELL READINGS (MV)	SPECIMEN NUMBER	FAILURE LOAD (KN)	HALF-CELL READINGS (MV)
PFA1	C1-1D	81.1	---	C1-1C	90.9	---
	C1-2D	87.4	-279	C1-2C	145.4	-259
	C1-3D	88.6	-276	C1-3C	>150	-269
PFA2	C2-1D	67.4	---	C2-1C	116.5	---
	C2-2D	71.4	-259	C2-2C	>150	-260
	C2-3D	80.5	-265	C2-3C	>150	-259
PFA3	C3-1D	70.8	---	C3-1C	128.3	---
	C3-2D	76.6	-280	C3-2C	136	-242
	C3-3D	78.5	-267	C3-3C	>150	-254
PC1 (w/c = 0.32)	C4-1D	79.9	---	C4-1C	>150	---
	C4-2D	72.8	-270	C4-2C	>150	-261
	C4-3D	76	-286	C4-3C	>150	-285
PC2 (w/c = 0.42)	C5-1D	59	---	C5-1C	>150	---
	C5-2D	65.7	-300	C5-2C	>150	249
	C5-3D	61.5	-301	C5-3C	>150	-248
HAC	C6-1D	64.4	---	C6-1C	120	---
	C6-2D	49.9	-303	C6-2C	>150	-290
	C6-3D	58.9	-321	C6-3C	>150	-258

High Alumina Cement

Although HAC concrete exhibited a high one-day compressive strength as compared with the other concretes, the half-cell potential results for all concrete specimens (cover thickness = 50 mm) show half-cell potentials of less than - 300 mV for PC concrete (w/c ratio = 0.42) and less than -500 mV for HAC concrete over the first 60 days. This suggests a high probability of corrosion compared with the other concretes for the same concrete cover thickness. However, with the passage of time (age = 150 days), the half-cell readings increased considerably (about - 100 mV), showing that the embedded steel had become passivated and it was less prone to corrosion.

Other Tests

Some other tests have been undertaken recently, such as the chloride ion profile Figure 7 along the concrete cover, petrographic tests, electron microscopic and x-ray diffraction tests to discern the products of corrosion, carbonation, permeability and porosity of the concrete. The results are being analysed presently and will be reported later.

SUMMARY AND CONCLUSIONS

The results of a preliminary investigation involving 14 direct tension specimens, subjected to seven different levels of acceleration, ranging from no corrosion to extensive corrosion, and electrochemical tests are reported. The level of corrosion was found to have a significant effect on bond at the steel-concrete interface. An 17.5 percent loss of cross-sectional area caused a 92 percent loss in the interface bond.

Partial results of a detailed study involving 192 lollipop (pullout) specimens with six different concrete mixes (three involving PFA), four different concrete covers (25, 50, 75 and 100 mm) and subjected to four different levels of corrosion, ranging from no corrosion to extensive corrosion, and many different electrochemical and other tests are reported. More results from this detailed study and an associated investigation on two abandoned bridges in Canada will be reported later.

ACKNOWLEDGEMENTS

The authors would sincerely like to thank the Natural Sciences and Engineering Research Council (NSERC Canada) for the Strategic Research Grant for this and other related studies. The authors would like to acknowledge the valuable contributions of the various industrial partners, and several individuals who contributed to the success of this project.

REFERENCES

1. ANDRADE, C., ALONSO, M.C., AND GONZALEZ, J.A. (1990), "An Initial Effort to Use the Corrosion Rate Measurements for Estimating Rebar Durability," Corrosion Rates of Steel in Concrete, Special Technical Publication 1065, American Society for Testing and Materials, Philadelphia, Pennsylvania, pp. 29-37.

2. AMLEH, L., AND MIRZA, M.S. (1998), "Corrosion Influence on Bond Between Steel and Concrete," Paper accepted for publication in the ACI Structural Journal.

3. AMLEH, L. (1996), "Influence of Corrosion of Reinforcing Bar on Bond Between Steel and Concrete," MEng. Thesis, McGill University, Montreal, Canada.

4. RASHEEDUZZAFAR, EHTESHAM, H.S., AND AL-SAADOUN S.S. (1992), "Effect of Tricalcium Aluminate Content of Cement on Chloride Binding and Corrosion of Reinforcing Steel in Concrete," ACI Materials Journal, pp. 3-13.

5. AL-SULAIMANI, G. J., KALEEMULLAH, M., BASUNBUL A., AND RASHEEDUZZAFAR (1990)," Influence of Corrosion and Cracking on Bond Behavior and Strength of Reinforced Concrete Member," ACI Structural Journal, Volume 87, No.2, pp. 220-230.

6. MALHOTRA, V.M. (1994), "CANMET Investigations Dealing With High Volume Fly Ash," Advances in Concrete Technology, Second Edition, Natural Resources Canada, Ottawa, Canada, pp. 445-482.

7. AMLEH, L. (1998), "PhD Research Proposal - Influence of Corrosion on Bond Between Steel and Plain and Fly Ash Concrete," Department of Civil Engineering and Applied Mechanics, McGill University, Montreal, Canada.

POLYMER-CONCRETE INTERACTIONS

N R Short
I M Shaw
Aston University
United Kingdom

ABSTRACT. Organic polymers are increasingly being used to modify the properties of cementitious materials. Although a significant amount of research has been conducted into the effect of modification on their mechanical properties, little is known about the nature of the interface and interactions taking place between the cement and polymer phases. This paper considers the case when the monomer methyl methacrylate (MMA) comes into contact with hydrated cements and is subsequently cured to give poly(methyl methacrylate) (PMMA). The techniques used to reveal the interfacial regions, and subsequently characterise them and their associated reaction products, are described. It was found that the monomeric MMA reacted with the cement to form primarily calcium methacrylate. This reaction proceeds in two stages (a) hydrolysis of MMA to methacrylic acid and (b) reaction of methacrylic acid with basic phases in cement, primarily calcium hydroxide. Water is a limiting reagent in these reactions. Since methacrylates are highly soluble they are likely to be detrimental to the properties of such composites. This emphasises the need for careful drying of the cement substrate prior to application of the monomer.

Keywords: Cement, Surfaces, Interfaces, Polymers, Interactions, Analysis, Durability.

Dr Neil R Short is a Senior Lecturer, Aston University. His current research interests are concerned with the durability of construction materials and include coatings for protection of metals in cementitious and acid rain environments, polymer modified cements, assessment of fire damaged concrete and durability of grc.

Dr Iain M Shaw The late Iain Shaw was a Research Student and then subsequently a Research Fellow at Aston University working in the field of polymer-cement composites.

INTRODUCTION

In recent years there has been increasing interest in the use of organic polymers to modify the properties of cement and concrete. Systems include polymer-concrete (PC), polymer modified cements (PMC) and polymer impregnated concrete (PIC). These polymer-cement composites are used as high performance materials in the construction and other industries, see for example reference [1].

Although a significant amount of data has been obtained on the mechanical and other properties of these composites, relatively little is known about the nature of the polymer-cement interface and interactions that may take place between the two components. In the wider field of concrete composites it has been recognised that interfaces between different components e.g. aggregate-cement or fibre-cement are important in determining properties such as strength and durability [2,3]. It is reasonable to assume that this will also be true in the case of polymer-cement composites.

A number of investigations have been initiated in our laboratories to provide information to help interpret properties of polymer-cement systems by understanding the nature of the interface between the two components. In this paper the approach taken and results obtained are reported for the case when the monomer methyl methacrylate (MMA) comes into contact with hydrated cements and is subsequently cured to give poly(methyl methacrylate) (PMMA). This system was chosen since it has found application, with somewhat variable success, for: (i) impregnation of hardened concrete e.g. bridge decks and more recently encapsulation of radioactive wastes [4], (ii) the matrix in polymer-concrete and (iii) the repair of concrete by crack injection.

In all three applications the concrete or aggregate should be dry prior to contact with the monomer and in examples (ii) and (iii) it is known that adhesion between the two phases is not very good if this is not the case [5]. Conversely, in example (i) attempts at extracting all the PMMA from impregnated cement has not proved successful [6]. These observations suggest that interactions are occurring between the cement and polymer phases. Thus a major aim of this work was to identify any possible reactions and their implications with regard to composite properties.

TECHNIQUES FOR INVESTIGATING INTERFACES

Studying the interface between the cement and polymer phases in a composite involves a number of practical difficulties. The actual interfacial region itself is hard to reveal and if interaction products have been formed, amounts may be very small and difficult to extract and analyse separately from the cement or polymer phases.

A number of specimen geometries were tested with the primary aim of achieving some mechanical interfacial failure thus permitting surface examination of the interfacial region. This proved difficult with the locus of failure often passing through bulk material. An alternative method is the selective dissolution of either the polymer or cement phase, in theory leaving the interfacial region available for surface analysis. In addition the amounts of interaction products may be magnified by increasing the surface area e.g. using a powdered cement rather than trying to cast against a flat surface. A wide range of instrumental methods of analysis are now available for investigations and these have been recently reviewed by a number of authors, see for example [3,7,8].

EXPERIMENTAL

Specimen Preparation

Specimens were made using Portland cement of class 42.5N to BS12 [9] for which the chemical composition, expressed in percentages by mass of the constituent oxides, is shown in Table 1. Samples of cement paste were made with w/c of 0.4 and 0.6, cast into cylindrical PVC moulds and compacted by means of vibration. The moulds were tightly sealed and subject to rotation to avoid segregation. After 24 hours the samples were de-moulded and cured at 25°C / 95% RH for periods up to 240 days. The hydrated cement was ground to pass a 300 µm sieve and dried at 105°C for 48 hours. To determine the influence of free water content of the cement, further samples were produced with 0, 25, 50 and 100% of the free water content.

Table 1   Main chemical properties of Portland cement (wt-%)

CaO	SiO_2	Al_2O_3	Fe_2O_3	SO_3	MgO	Na_2O	K_2O	LOI*
64.5	20.0	5.3	3.4	3.0	1.2	0.1	0.8	0.9

* Loss on ignition

To make polymer cement composites, 5g of hydrated cement powder was added to an equal mass of MMA which contained benzoyl peroxide (BP) (50:1 by mass) as initiator. The samples were heated to 65°C with stirring, in order to avoid segregation. When the mixture began to increase in viscosity the temperature was increased to 100°C for one hour to ensure maximum polymerisation. Alternatively 5 g of hydrated cement powder was refluxed with MMA (no initiator present) for one hour. The mixture was allowed to cool and most of the MMA decanted off whilst the remainder was removed by evaporation. The residue was then subjected to chemical analysis. This method was used to provide the most favourable conditions for any potential reactions to take place. In addition individual samples of cement, polymer and potential reaction products were prepared as standards. The preparation procedures are described in greater detail in reference [10].

Analytical Techniques

When using DTA to investigate polymer-cement reactions there is a concern that the technique may give misleading information [11]. This is because reactions may take place during the DTA run itself, either masking reactions which have taken place prior to thermal analysis or implying that reactions may have taken place which in fact did not occur.

To eliminate this problem both samples of polymer cement and equivalent physical mixtures were investigated. Specimens were filed and sieved to 300 μm and examined using differential thermal analysis (DTA). Equal mass of sample was used in each run in order to aid in the interpretation and comparison of the results.

For the infrared spectroscopy (IRS) a few milligrams of the sample were mixed with KBr powder and pressed into a thin disc. The disc was placed in the sample holder of a Perkin Elmer 1710 fourier transform infrared spectrometer. The instrument scanned the 4000 to 220 cm^{-1} region of the infrared, and absorptions were detected at specific, well defined, wavelengths [10].

Ion chromatography proved to be very valuable in identifying soluble ions and was carried out using a Dionex 2000i/SP instrument. An AS6 anion column was used for the separation of ions, with 0.001 mol l^{-1} $NaHCO_3$ as the eluent, at a flow rate of 0.1 ml min^{-1}. Data was recorded and digitised so as to calculate concentrations from peak areas. Water extracts were prepared by shaking 4g of sample, sieved to 75 μm, with 100 ml of distilled water for 30 min, then filtered. The filtrate was made up to 250 ml, this needing no further dilution for injecting into the ion chromatography column [12].

Sample surfaces were analysed by X-ray photoelectron spectroscopy (XPS) with Mg Kα radiation as the excitation source. General information was obtained from wide scan survey spectra, whilst more detailed knowledge was obtained from high resolution spectra of C_{1s}, K_{2p}, O_{1s}, $Ca_{2p3/2}$, and $Si_{2p3/2}$ peaks [13]. This technique was primarily used to determine the influence of drying on the surface chemistry of cements and to try and identify interaction products remaining after selective dissolution of the polymer.

RESULTS AND DISCUSSION

DTA

The thermograms for physical mixes and polymer cements are shown in Figures 1(a) and (b), respectively, and are representative of several repeated runs. Endotherms and exotherms were identified by reference to the thermograms obtained from standards of the individual components i.e hydrated cement, MMA, PMMA and calcium methacrylate. In the case of the physical mix, Figure 1(a), there is an endotherm at around 340°C followed by an exotherm at 460°C. These may be attributed to decomposition and oxidation of the PMMA. The well defined endotherm at 530°C is from the decomposition of calcium hydroxide.

The small endotherm at around 700°C suggests the presence of carbonate, which was absent in the PC standard, and suggests that an interaction between the polymer and cement during the DTA run itself has occurred. In the case of polymer cements, Figure 1(b) the well-defined calcium hydroxide peaks at 530°C, are still present but significantly diminished compared to those found in the case of the physical mixes. This difference may be seen as evidence to support the hypothesis that there is some reaction between calcium hydroxide and MMA in the polymer cements. The peaks were too small for quantitative measurement, but the diminution seems quite clear and was reproducible.

Additionally the endotherm at 420°C may be attributed to the presence of calcium methacrylate, since a similar one was observed in its standard.

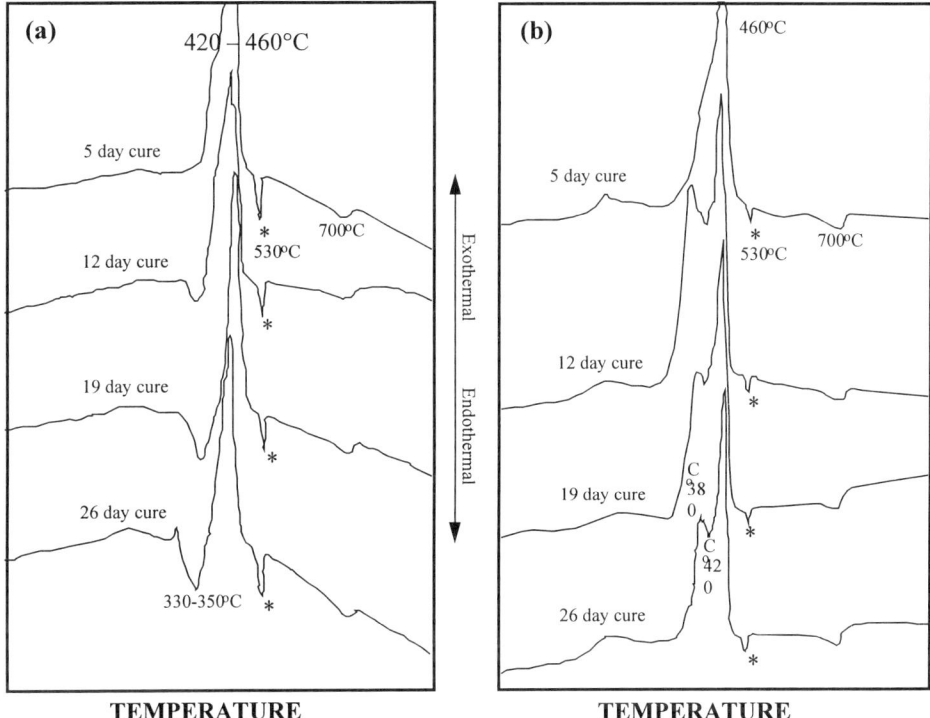

Figure 1 DTA thermograms of (a) physical mixes (b) polymer cements
(* Note diminution of the calcium hydroxide peaks in polymer cements compared to physical mixes)

IRS

In the IRS analyses, with both polymer cements, Figure 2(a) and physical mixes, Figure 2(b), it was easy to identify absorptions owing to cement and polymer, following earlier characterisation of the individual phases. At first sight there is no obvious difference between polymer cement the physical mixes. However, closer observation shows that there are some differences particularly a sharp absorption present at 1601 cm^{-1} which is evident in the polymer cement but absent in the physical mix. This was the case for all samples examined, the difference being most noticeable by comparison with the adjacent absorption band at 1645 cm^{-1}. Identification of the 1601 cm^{-1} band is of great importance since there are only a few possibilities for it, the presence of carboxylate groups being the most likely [14].

The consumption of calcium hydroxide as shown by DTA in Figure 1, to leave a reaction product containing carboxylate groups as shown by IRS in Figure 2, supports a hypothesis that MMA is hydrolised to methacrylic acid in the highly alkaline conditions found associated with cements, the latter reacting immediately with calcium hydroxide to form calcium methacrylate.

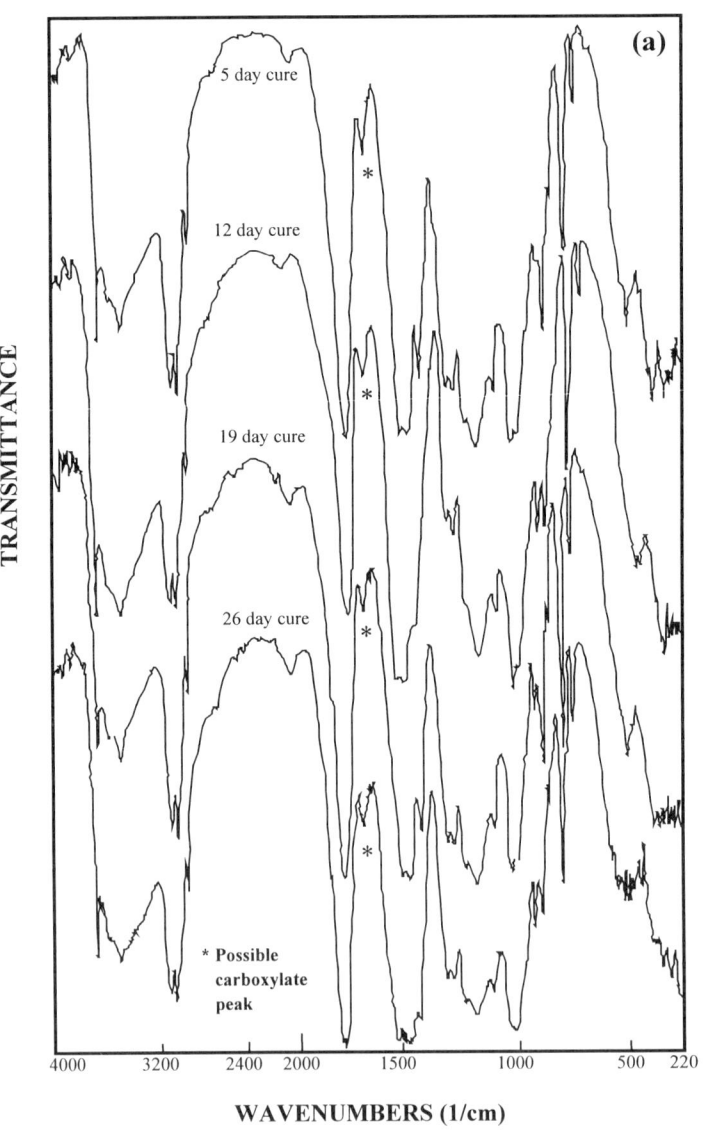

Figure 2(a) Infrared spectra of physical mixes

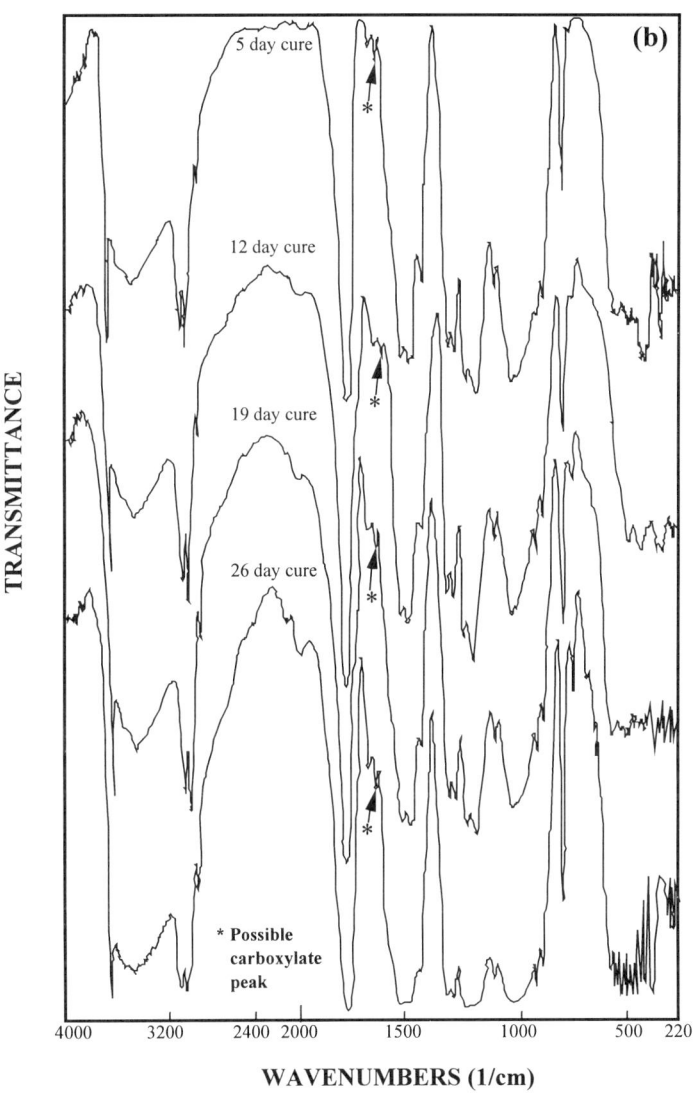

Figure 2 (b) Infrared spectra of polymer cements

Ion Chromatography

Only anions were of interest, since reaction products of the polymer would most likely be negatively charged methacrylates. Mixed sodium salts of methacrylate, benzoate, chloride, and sulfate were tested as standards and these produced quite distinct peaks in the chromatagrams as can be seen in Figure 3(a). The elution order was benzoate, methacrylate, chloride and sulfate at elution times of around 2, 6, 9 and 14 minutes respectively.

Extracts from the cement used, Figure 3(b), showed four clearly defined peaks at elution times of around 3, 4.5, 5.5 and 12 minutes. These were not conclusively identified but probably arise from aluminates, sulphoaluminates or aluminosilicates. The significant factor is that the methacrylate peak is easily resolved in the presence of the most likely other ions to be found in polymer-cements. Additional experiments showed that traces were not dependent on W/C or curing times used.

Figures 3(c) and (d) show the chromatograms of solutions extracted from the residue after refluxing and a polymer-cement composite respectively. As with the cement extracts both chromatograms showed the peaks attributable to cement ions with an extra peak at an elution time of around 6 minutes attributable to methacrylate ion. The presence of methacrylate is evidence of reaction between the methyl methacrylate and cement.

The influence of free water content of the cement on the concentration of methacrylate produced is shown in Figure 4. It can be seen that the amount of methacrylate formed is proportional to the initial amount of water present in the cement.

That methacrylate formation has occurred in both the refluxed and polymer cements suggests that reaction is with the monomer and not the polymer. Thus in PC or PIC reaction will occur either prior to or during polymerisation and will be independent of it.

Nature of the Interactions

From the observations using DTA, IRS and ion chromatography it may be deduced that the formation of methacrylate salts proceeds through two stages:

1. This involves hydrolysis of MMA in the highly alkaline environment to give methacrylic acid and methanol. Water is a limiting reagent in this reaction.

$$H_2C=C(CH_3)C=O_2CH_3 + H_2O \rightarrow H_2C=C(CH_3)C=O_2H + CH_3OH \qquad (1)$$
methyl methacrylate — methacrylic acid — methanol

2. Subsequent reaction of the methacrylic acid occurs with basic cement minerals, particularly calcium hydroxide, to give calcium methacrylate.

$$2\{H_2C=C(CH_3)C=O_2H\} + Ca(OH)_2 \rightarrow \{H_2C=C(CH_3)C=O_2\}_2Ca + 2H_2O \qquad (2)$$
methacrylic acid — calcium methacrylate

Since the calcium methacrylate formed is highly water soluble a reaction such as this would be generally detrimental to the properties of a polymer - cement composite.

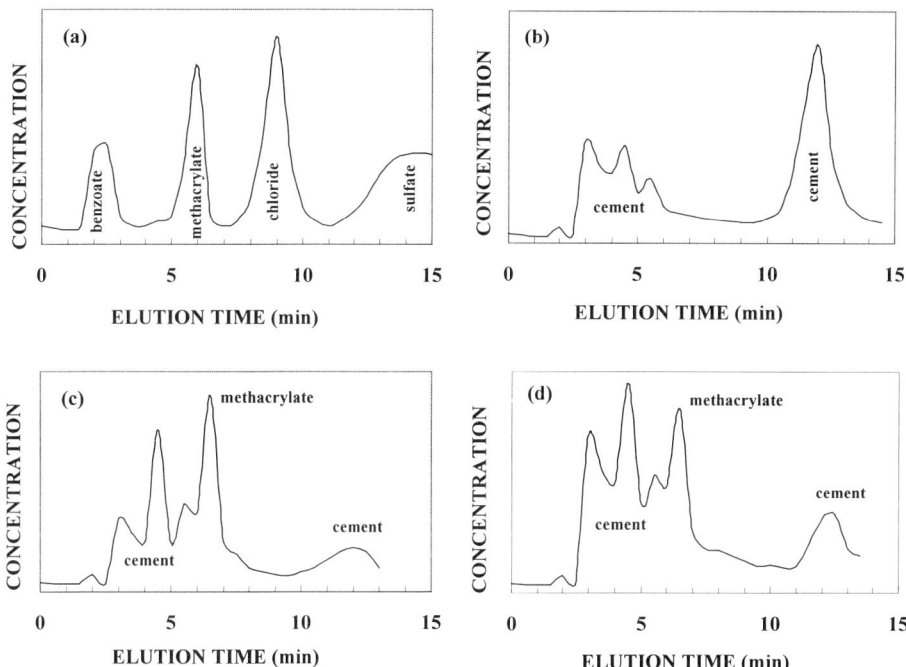

Figure 3 Ion chromatograms showing: (a) mixed sodium standards, and ions extracted from: (b) hardened cement paste, (c) hardened cement paste refluxed with methyl methacrylate, (d) polymer cement, **Note:** Sensitivity adjusted to give maximum peaks

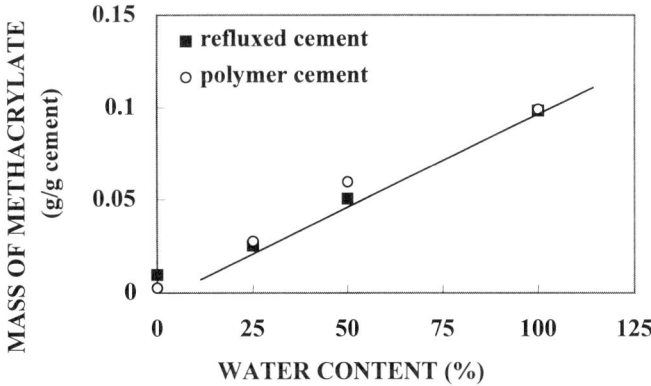

Figure 4 Mass of methacrylate (g/g cement) versus the water content of cement as a % of the water before drying at 105°C

Influence of Drying on Surface Chemistry

A number of cement surfaces were examined using XPS, the samples having been prepared with different w/c (0.4, 0.7), cure time (28, 90 days) and with fractured or machined surfaces. No significant peak shifts or relative height variations for the elements examined were observed, indicating identical chemical environments. This suggests that over the areas irradiated (~100 mm^2) any chemical differences between different minerals become averaged to produce surfaces homogenous on this scale.

Drying specimens by heating at 105°C, placing in a vacuum or alcohol displacement resulted in little changes in the position and size of any peaks other than that from potassium. It was found that the K_{2p} peak was enhanced at the surface of heated specimens by over twice that found within the bulk. This was not so for specimens dried in a vacuum or by alcohol displacement. It would seem that since KOH is highly soluble in water, particularly at elevated temperatures, potassium ions are transported through the pore system from within the specimen to the evaporation surface. This is not found with alcohol displacement since KOH is insoluble in isopropanol. With vacuum drying either the drying times are slower and solubility of KOH in cold water is lower or water transportation occurs by vapour phase. Thus the surfaces of specimens dried by heating probably possess a higher alkalinity compared to other surfaces.

Interfacial Reaction Products

Cements with PMMA bonded to them had the polymer removed by dissolution with chloroform. This should leave ionically bonded cement/polymer reaction products on the revealed cement surface. Compared with cement surfaces differences in surface chemistry were noted, in particular the abundance of carbon on surfaces exposed to MMA. Although several of the carbon peaks could be identified e.g. contamination was present on all specimens, the nature of others was uncertain. Nevertheless, it was concluded that a significant amount of carbon compounds were still adhering to the cement surface after treatment with a good solvent. Whilst these were interpreted as being reaction products it has not been possible to identify them. However, it is possible that methacrylic acid molecules become incorporated in the growing polymer chains which could then bond very strongly with the cement minerals when moistened. Subsequently this would be expected to improve the mechanical properties of the composite.

PRACTICAL APPLICATIONS OF RESULTS

It is important to study interactions between organic polymers and cements in order to aid in the selection of compatible systems which confer desirable properties whilst avoiding detrimental reactions. To obtain a clear understanding of interfacial behaviour and avoid introducing artifacts of the methods used, it is advisable to approach the problem from different perspectives and use as wide a range of techniques as possible. Thus whilst polymer cement interfaces are difficult to reveal mechanically they can be exposed if suitable solvents for the polymer (or cement) are used. In this state, reaction products may then be studied by techniques of surface analysis.

Whilst this approach may not always be completely successful useful information can still be obtained. Similarly contact surface areas may be increased by using a powdered substrate although this will need to be taken into account when interpreting results. It should be remembered that the chemistry of pore surfaces (and their constricted nature) may well result in different reactions compared with reactions at a powdered hydrated cement surface, e.g. it may influence the kinetics of polymerisation.

Using a range of techniques it has been possible to show a reaction can occur between MMA and cement provided that moisture is present. In practice, if the highly water-soluble methacrylate were leached from the polymer/cement interface over time then the properties of the composite would be modified. This confirms the need for thorough drying of the cement or aggregate prior to contact with the monomer if adhesion is to be maximised.

Improvements in bonding have been made via the use of coupling agents, although again variable success has been reported. The techniques employed here could be used to determine the most effective systems and their tolerance to the presence of moisture.

CONCLUSIONS

1. To investigate polymer-cement interactions it has been found useful to employ a range of techniques and analytical methods.

2. Methyl methacrylate reacts with cements to form methacrylates, primarily calcium methacrylate. This reaction is independent of the polymerisation reaction. These reactions proceed through two stages (a) hydrolysis of MMA to methacrylic acid and (b) reaction of methacrylic acid with basic phases in cement primarily calcium hydroxide.

3. Water is a limiting reagent in the above reactions. If water is present then such reactions are likely to be detrimental to the properties of polymer-cement composites since the phases formed are water-soluble. Thus from a practical point of view samples should be thoroughly dried prior to contact with the monomer.

REFERENCES

1. POLYMERS IN CONCRETE. Proc. 2nd East Asia Symposium, Eds. Y Ohama, M.Kawakami and K Fukuzawa, E & FN Spon, 1997, pp 539.

2. INTERFACES IN CEMENTITIOUS COMPOSITES. Proc. RILEM Int. Conf., Ed. J C Maso, E & FN Spon, 1993, pp 310.

3. ROY D M AND JIANG W. Influences of interfacial properties on high-performance concrete composites. Mat. Res. Soc. Symp. Proc. 1995, Vol. 370, pp 309-318.

4. ISHIZAKI K. Fundamental properties of PIC and its application in containers for conditioning and disposal of low radioactive wastes. Proc MATETA workshop on high Flexural polymer-cement composites, Japan, 1996, pp 29-41.

5. RODLER D J, WHITNEY D P, FOWLER D W AND WHEAT D L. Repair of cracked concrete with high molecular weight methacrylate monomers. ACI SP 116-7, 1988, pp113-127.

6. OHAMA Y. Molecular weight of polymer formed in polymer-impregnated concrete. Proc 1st Int. Conf. on Polymers in Concrete, 1975, pp 60-63.

7. SCHULSON E M. Ice damage to concrete, US Army Corps of Engineers, Special Report 98-6, April 1998, pp 29-43.

8. POPOOLA O, KRIVEN W M AND YOUNG J F. High-resolution electron microscopy and microchemical characterisation of a polyvinyl alcohol acetate / calcium aluminate composite (macro-defect free cement). Ultramicroscopy, Vol. 37, 1991, pp318-325.

9. BRITISH STANDARDS INSTITUTION. Specification for Portland cement. BS12:1996.

10. SHAW I M. Interactions between organic polymers and cement hydration products. PhD Thesis, Aston University, 1989.

11. BEN-DOR L. Thermal Analysis. In: Advances in Cement Technology. Ed. S N Ghosh, Pergamon Press, 1983, pp695-696.

12. DIONEX CORPORATION. Dionex ion chromatography cookbook, a practical guide to quantitative analysis by ion chromatography. 1987.

13. WATTS J F. An introduction to surface analysis by electron spectroscopy. Oxford University Press, 1990.

14. BELLAMY L J. The infrared spectra of complex organic molecules. Chapman and Hall, London (1975).

UNSATURATED MASS TRANSFER WITHIN HARDENED CEMENT PASTE MODIFIED BY THE PARTIAL REPLACEMENT WITH SILICA FUME

D Fairhurst

A Platten

University of Central Lancashire

United Kingdom

ABSTRACT. In this paper the concept of the Sorptivity is appraised as a parameter relating material structure to the transfer of moisture in the unsaturated condition. Published values for the Sorptivity of hardened cement pastes are scarce for silica-fume blended pastes. This paper presents values of the Sorptivity for pastes at 1, 3 and 7 days of age for 5 and 10% replacements. Values are presented for the vapour phase permeability obtained by diffusion cell technique, of 1, 3, and 7 day old hardened cement pastes, modified by partial replacement with silica-fume at 5% and 10% levels. Early age characteristics are considered with regard to the use of silica fume modified cement pastes as a repair system for existing structures.

Keywords: Silica fume, Water sorption, Porosity, Vapour transfer, Diffusivity

Don Fairhurst is a full time research student working in the Department of Built Environment at the University of Central Lancashire. He is nearing the completion of his thesis to investigate the barrier properties of cement pastes which are modified by silica fume.

Dr Andrew Platten is a principal lecturer at the University. His research background has related to mass transfer characteristics of porous building materials. This work has included both moisture sorption and desorption processes in the liquid and vapour phase. His work has involved applications to ambient conditions and conditions at elevated temperatures. He has a keen interest in the development of computer modelling of such phenomena.

INTRODUCTION

The fundamental transport mechanism underlying the unsaturated mass transfer through a porous system is the capillary diffusion. If potentials which arise from temperature gradients or differences in salt concentration are assumed to be negligible, the hydraulic capillary Diffusivity can be viewed as a dominant influence upon of the moisture content of the porous system. Under these defined conditions, the movement of moisture within the porous system is due to potentials arising from moisture gradients within the system. For practical purposes, the unsaturated state more readily reflects field conditions for building structures which are subject to wetting and drying cycles. Such environments seldom subject the structure to conditions of the entire pore volume being saturated. Saturated flow is thus "...rare, occurring only in weak concretes or under very great external heads" [1].

This places an emphasis upon the study of unsaturated mass transfer and the need to develop models that will allow predictive analysis of the serviceability of cementitious structures. The development of such models require that the components of unsaturated flow be isolated and determined experimentally. To facilitate this, three transfer mechanisms have been studied for hardened cement pastes partially replaced by pozzolanas. Dispersion effects associated with the presence of aggregates within a concrete have been ignored; therefore the values and models presented are based on the structural properties of the pastes themselves.

In this paper, the Sorptivity is discussed with values presented for the Sorptivity of Condensed Silica-fume (csf) pastes. Unsaturated flow theory, however describes two phases of moisture movement within the porous body (liquid and vapour), therefore values for the vapour Permeability are also presented in this paper.

GENERAL CONCEPTS

Derivation of the Sorptivity and General Unsaturated Flow Equation

It is common place to consider the parameter of Sorptivity as the rate of hydraulic absorption and infiltration of moisture into the porous matrix. The Sorptivity as a material property is related to the mass transfer of moisture within an initially dry porous body, where the driving force is predominantly derived from capillary tension head.

Philip [2], in the analysis of the absorption of moisture in the unsaturated condition within porous media described the cumulative absorption (i), with respect to time (t) and distance (x). In this instance t and x, are expressed as a single function of λ:

$$\lambda(x,t) = x\, t^{-0.5} \qquad (1)$$

Thus the cumulative mass absorbed, i for elapsed time is described as:

$$i = \left[(\theta_1 - \theta_0)\int_0^\infty \lambda\, d\theta\right] \cdot t^{0.5} \qquad (2)$$

The cumulative absorption is related to a material parameter referred to as the Sorptivity, where:

$$S = (\theta_1 - \theta_0) \int_0^d \lambda \, d\theta \tag{3}$$

Equation (1) denotes the relationship for the distance advanced for the infiltrating moisture as linear with respect to the square root of time. Thus, not only can the cumulative mass absorbed be predicted, but also the moisture content and its position in the porous body.

Experimentally, the moisture content and its position can be found either destructively [3], or by non-destructive means such as Gamma-spectrometry [4], and Electrical Resistivity [5]. A simplified solution exists [3], where if the mass transfer properties are known then prediction of the both the cumulative mass absorbed and distance advanced of the wetting front is possible. Though these solutions are an over simplification, more readily suited to soil hydrology, the basic relationship of porosity to both absorption and distance advanced are shown:

$$i = \left(-2 n_e \, k \, \Delta h_t \, A^2\right)^{0.5} \cdot t^{0.5} \tag{4}$$

and

$$x = \left(\frac{-2 k \, \Delta h_t}{n_e}\right)^{0.5} \cdot t^{0.5} \tag{5}$$

where,
k = the permeability (m sec^{-1})
h_t = the hydraulic suction derived from the internal pore structure (m)
n_e = the effective porosity

It is clear that i is proportional to the square root of the effective porosity, n_e and that x is inversely proportional to the square root of n_e. The relationship of $i(t^{0.5})$ is considered in this paper.

Equations (4) and (5) introduce the components of Darcy's saturated flow equation, where:

$$i = -k\left(\frac{dh}{dx}\right) \tag{6}$$

The term dh/dx represents the hydraulic potential. This expression is valid for flow both in the saturated unsaturated conditions, where k is a function of θ which exhibits high values at water contents at or close to saturation and decreasing to zero at dryness. The reduction in k with reducing θ is primarily due to reductions in the cross section of pore space available for flow and a reduction in the continuity of the liquid phase. Thus flow must occur by other means than direct transfer [6]. Further and h_t is a function of θ whose magnitude is greater at low values of θ and zero at saturation. Thus, k and h are qualified as functions of the moisture content, $k(\theta)$ and $h(\theta)$ with the result that the extended Darcy flow equation for unsaturated flow is expressed as:

$$\frac{\partial \theta}{\partial t} = -\frac{\partial}{\partial x} k(\theta) \left(\frac{\partial h}{\partial x} \right) \tag{7}$$

or, in the form of a moisture concentration difference, by:

$$\frac{\partial \theta}{\partial t} = \frac{\partial}{\partial x} \left(D(\theta) \frac{\partial \theta}{\partial x} \right) \tag{8}$$

where,

D = hydraulic Diffusivity (m² s⁻¹)

and is defined by:

$$D(\theta) = K(\theta) \frac{\partial \psi}{\partial \theta} \tag{9}$$

where,

ψ = capillary force (m) equivalent to ht

Equations (7) and (9) form the basis for deriving solutions to the vapour phase flow are now discussed.

Vapour Phase Unsaturated Flow

In the case of vapour transfer, ψ is the moisture potential in thermodynamic equilibrium with the water in a porous material with atmospheric pressure as datum, and is related to vapour pressure by:

$$rh = \exp\left(\frac{-Mg\psi}{RT} \right) \tag{10}$$

thus:

$$\left(\frac{RT}{Mg} \right) \ln rh = -\psi \tag{11}$$

Thus, where the vapour pressure potential determined by Equation 11, the vapour permeability is denoted as:

$$i = -k_v \frac{\Delta \psi}{x} \tag{12}$$

with the vapour Diffusivity solved using Equation 9 by:

$$D_v(\theta) = k_v(\theta) \left(\frac{\partial \psi}{\partial \theta} \right) = k_v(\theta) \frac{\left(\frac{RT}{Mg} \right) \ln rh}{\partial \theta} \tag{13}$$

EXPERIMENTAL WORK

The determination of the Sorptivity was undertaken using the method detailed by Hall [7]; where the cumulative mass absorbed for an isotropic body, which exhibited an initially uniform moisture content was obtained with respect to time. Water was allowed to infiltrate vertically into the sample from a free-water reservoir.

Vapour phase analysis was undertaken using a modified method of BS 3177 (1959). The samples were subject to various relative humidities in a Vindon humidity cabinet with humidity levels produced by the saturated salt concentrations given in Table 1. A vapour sink was obtained by the use of silica-gel which maintained zero moisture content within the diffusion cell.

Table 1 Saturated salt solutions and operating temperatures used in this study

	TEMPERATURE		
	20°C	30°C	40°C
Saturated Salt Solution	Relative Humidity (%)		
Sodium Chloride	76	**75**	75
Sodium Dichromate	55	**52**	50
Potassium Carbonate	44	**43**	42
Magnesium Chloride	33	**33**	32
Potassium Acetate	23	**22**	20
Lithium Chloride	12	**12**	11

* Values in **Bold** are the operating temperatures used in this work.

MATERIAL DETAILS

In this study, pastes used for the determination of the Sorptivity were modified by Elkem 500S "Microsilica", and for vapour phase analysis modified by "Micropoz". The cement pastes were produced by direct replacement of cement by 5% and 10% silica fume. The water cement ratio used for each specimen was 0.5. The mix process was undertaken by hand.

Samples used in the study of the Sorptivity were cast from 100mm diameter moulds, de-moulded after one day, and bath cured until the age of test. The samples were then cut to expose the structure of the paste and dried to a constant mass in a vacuum oven at $40^0 C$. Samples used in the study of vapour phase flow were cast from rings 40mm diameter ($\approx$2mm thickness), bath cured until the age of test and dried as before to a constant mass.

A constant water/cement ratio of 0.5 was used throughout the study.

RESULTS

Sorptivity

The Sorptivity for the pastes studied in this work has been derived from the gradient of a straight line plot of $i_{(t^{0.5})}$. For a straight line to be represented, four conditions must be satisfied. These are: 1) the material must be homogeneous throughout the region of ingress, 2) the capillary absorption must be normal to the inflow face, neither converging nor diverging, 3) the water must be freely available at the suction surface, and 4) gravitational effects must not affect the absorption process. The pastes studied in this work satisfy the four conditions.

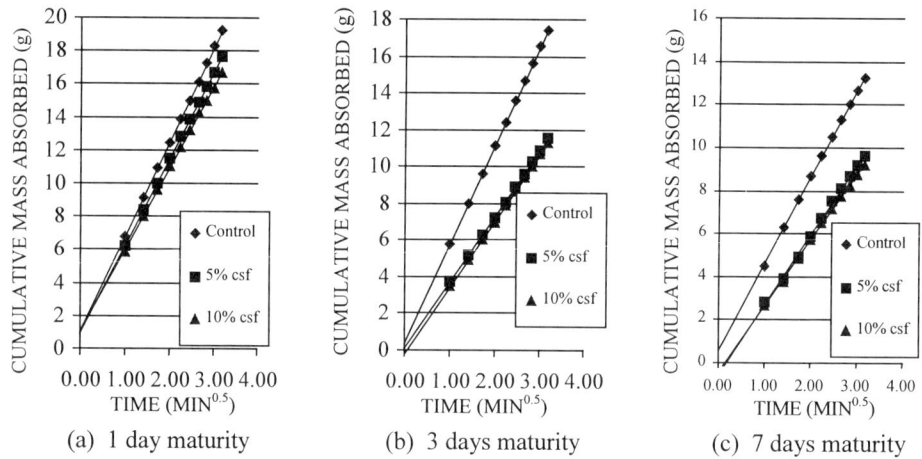

(a) 1 day maturity   (b) 3 days maturity   (c) 7 days maturity

Figure 1 Cumulative mass absorbed for elapsed time for pastes at 1, 3 and 7 days maturity

Qualification of the linear relationship of mass absorbed to the square root of time is illustrated in Figure 1, and therefore satisfied the four conditions of derivation from the gradient of $i_{(t^{0.5})}$. Collectively, the data illustrate the effects of hydration on the water absorption properties of the pastes. The data provides clear evidence of the effect of csf replacement, both for age of paste and percentage of replacement. Values for the Sorptivity of the pastes are presented in Table 2.

With regard to the data obtained the following observations can be made:

From Figure 1(a) at one day maturity, although the water absorption of the pastes are comparable, both 5% and 10% csf replacement pastes exhibit reductions in the order of 7% and 12% respectively compared to the control paste.

With respect to Figure 1(b) at 3 days maturity, the pastes exhibit reductions to 1 day values in the order of 6%, 32% and 27% for the control, 5% and 10% pastes respectively.

Further, comparison with the control paste for the same age shows that samples with 5% and 10% replacement to have lower Sorptivity values in the order of 30% and 32% respectively, where a distinct banding is evident for csf pastes.

Table 2 Values for the sorptivity

AGE (DAYS)	REPLACEMENT (%)	SORPTIVITY (mm MIN$^{0.5}$)
1	0	7.35
	5	6.82
	10	6.40
3	0	6.88
	5	4.62
	10	4.69
7	0	5.20
	5	4.20
	10	3.96

Finally, from Figure 1(c) at 7 days maturity reductions in the Sorptivity are exhibited by all pastes, with values representing reductions in the order of 29%, 38% and 38% from day 1 values. Comparison with the control paste at the same age, shows the csf pastes again exhibiting lower values, in the order of 32% and 36%, for 5% and 10% replacement.

For all ages of pastes, it is evident from the values presented that whilst there is some reduction of the Sorptivity for increased csf replacement, this reduction is negligible.

Relationship Between the Sorptivity and Porosity

The porosities of the pastes studied in this work are presented in Table 3. It is evident that at day 1, the porosities for all pastes are comparable and that at later maturities distinct reductions relative to csf replacement are apparent.

Table 3 Values for the porosity

AGE (DAYS)	REPLACEMENT (%)	POROSITY (n_e)	REDUCTION COMPARED CONTROL PASTE AT SAME AGE	REDUCTION TO VALUE AT DAY 1
1	0	40.94	-	-
	5	39.89	3%	-
	10	38.27	7%	-
3	0	35.36	-	14%
	5	32.71	7%	18%
	10	29.20	17%	24%
7	0	29.46	-	24%
	5	23.13	22%	42%
	10	15.85	46%	59%

A simplified method for deriving the cumulative mass absorbed has been discussed, Equation 4, where the absorption was shown to be proportional to the square root of the porosity. Whilst it is reasonable to assume the Sorptivity to be proportional to the porosity mathematically, the linear relationship is not found to exist in the case of the cement pastes studied in this work.

The relationship between sorptivity and porosity is illustrated in Figure 2. The plotted data follows a best fitted exponential curve (Sorptivity = $\exp(1.61x)$), though the accuracy of the fit is only 0.82 which is not sufficient to derive a confident relationship. It seems probable that for the case of cement pastes, the relationship is not linear due to the large pore size distribution and the effects of tortuosity, although this observation requires further investigation.

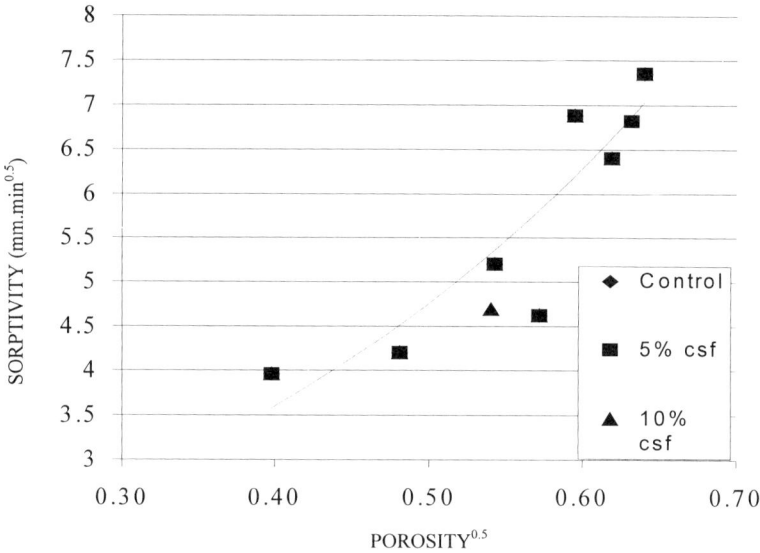

Figure 2   Relationship between the sorptivity and square root of porosity

Vapour Permeability

The vapour permeabilities of the cement pastes studied in this work illustrate reductions for increasing degree of hydration (age of paste at test), and for csf replacement. The latter is clearly shown in Figure 3a, where clear separation is evident between the control paste and the csf replacements. This trend is less pronounced at day 3 and 7, where all pastes exhibit comparable values.

All data illustrate that for increasing moisture content: i.e. relative humidity; the vapour permeability increases in magnitude. This relationship supports the proportionality of the mass transfer to the permeability as a function of the moisture content of the paste, as reported by Rose [6].

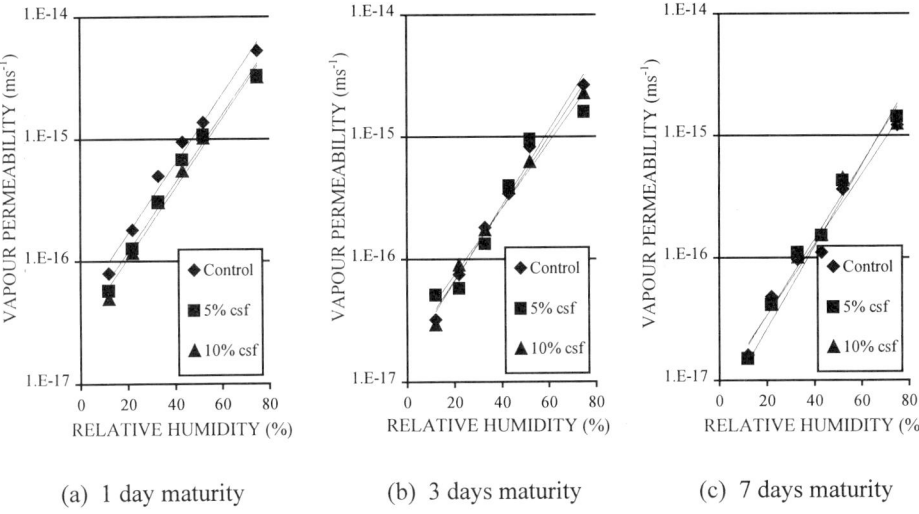

Figure 3 Vapour permeabilities with respect to relative humidity and age of pastes at 1, 3 and 7 days of age (3a, 3b, 3c)

CONCLUSIONS

The results presented in this paper indicate that csf replacement reduces the water absorption of cement pastes, both for increased percentage of replacement and for degree of hydration: i.e. maturity of the paste. The reduction in Sorptivity is most likely resultant to changes in tortuosity, capillary porosity and reducing capillary pore radii, which are associated with hydration and csf replacement, as reported previously by the authors [10] and by Pomery [11]. The Sorptivity can therefore provide qualitative information with regard to the potential of the paste to absorb moisture. Further in repair applications silica modified pastes indicate improved performance with respect to the resistance to moisture absorption and associated mechanisms of deterioration.

The proposition that the sorptivity is directly proportional to the square root of the porosity has been shown to be invalid in this instance. A tentative relationship may be considered, however. The data provides evidence of reduced values of the Sorptivity of 5% and 10% inclusions of micro silica.

A mathematical derivation of the unsaturated vapour transport mechanisms of permeability and diffusion has been presented based upon the liquid phase unsaturated flow equation. The vapour permeability of csf replacement pastes has been shown to be reduced at day 1, but thereafter (up to 7 days of age), values are comparable to normal paste.

The data provides evidence that the early age performance of modified pastes can be enhanced with respect to resistance to the ingress of water in the vapour phase. This observation may be used in support for such systems in repair applications.

Further, the vapour permeability has been shown to be a function of the moisture content and to increase in magnitude with increasing relative humidity.

In progression to this work it is recommended that studies are undertaken to examine the water content profiles in the early age condition for similar samples subject to the absorption of moisture. This study is recommended as a validation of the relationship defined in Equation 5. Further, studies concerning pore size and its relationship to the Sorptivity are advised for csf modified cement pastes.

ACKNOWLEDGEMENTS

Elkem Materials for supplying Elkem 500S and Phillip Jones for supplying the "Micropoz" csf. Stuart Pepper and Castle Cement for preparing the Micropoz into a slurrified form. Pam Curwen for her assistance in the experimental procedure.

REFERENCES

1. DUNAGAN, W H. Methods For Measuring The Passage Of Water Through Concrete, Proceedings of the American society for Testing Materials. Vol. 39, 1939, pp. 866-880.

2. PHILIP, J R. The theory of infiltration: 4. Sorptivity and algebraic infiltration equations, Soil Science, 84, pp. 257-264, 1957.

3. KIRKHAM, D, and POWERS, W L. Advanced soil physics, Wiley Interscience, New York, 1972.

4. KUMARAN, M K, and BOMBERG, M. A Gamma-spectrometer for determination of density distribution and moisture distribution in building materials, Proc. Int. Symp. Moisture and Humidity, 1985, pp. 485-490.

5. M^{C}CARTER, W J, CHRISP, T M, EZIRIM, H, and BASHEER, P A M. In-situ evaluation of properties of concrete in the cover zone, Proceedings of the International Congress Concrete In The Service Of Mankind. Concrete Repair, Rehabilitation and Protection. Ed Dhir and Jones. E & FN Spon, 1996, pp.113-123.

6. ROSE, D A. Water movement in porous materials: Part 2-The separation of the components of water movement, Brit. J. Appl. Phys., 1963, VOL. 14.

7. HALL, C. Water Sorptivity of mortars and concretes: a review, Magazine of Concrete Research, Vol. 41, No. 147, June 1989, pp. 51-61.

8. HALL, C, HOFF, W D, NIXON, M R. Water Movement in Porous Building Materials - VI. Evaporation and Drying in Brick and Block Materials, Building and Environment, Vol. 19, No. 1, pp. 13 - 20, 1984.

9. EMERSON, M. Mechanisms of Water Absorption by Concrete, Proceedings of the International Conference on Protection of Concrete. Dundee, Scotland. pp. 11-13, September, 1990

10. FAIRHURST, J and PLATTEN, A K. Early Age Observations by SEM of the Changes to Hardened Cement Paste with the Partial Replacement by Silica-fume. Concrete In The Service Of Mankind. Concrete for Infrastructure and Utilities. Ed Dhir and Henderson. E & FN Spon, 1996. pp. 63 – 72.

11. POMEROY. Setting and early strength gain of cement containing fly ash, slag or limestone fillers. Testing During Concrete Construction. Ed Reinhardt. Chapman and Hall. 1990.

HYGRIC MATERIAL PROPERTIES OF MECHANICALLY LOADED CONCRETE

J Drchalova S Hoskova
J Toman T Klecka
R Cerny
Czech Technical University
P Jurek
Institute of Physics
Czech Republic

ABSTRACT. The water vapour permeability, δ, and liquid moisture diffusivity, κ of cement mortar are measured on both unloaded samples and samples mechanically loaded up to 90% of compressive strength. Over a relatively very wide range of load, all values of δ and κ are found to be within the errorbar of experimental methods employed. Significant differences compared to the unloaded state are observed only at 90% of compressive strength. Scanning electron microscope images show that the main reason for these differences is the appearance of cracks with a typical width of 1 - 2 μm.

Keywords: Moisture diffusivity, Water vapour permeability, Mechanical load.

Dr J Drchalova and Dr S Hoskova are Assistant Professors at the Department of Physics, FCE CTU Prague. They specialize in measuring hygric properties of building materials.

Professor J Toman is Professor of Physics at the Department of Physics, FCE CTU Prague. His main research interest is in measuring thermal properties of building materials under non-standard conditions.

Dr T Klecka is Director of Klokner Institute, CTU Prague. He specializes in measuring mechanical properties and analysing the porous structures of building materials.

Dr P Jurek is a Research Worker in the Institute of Physics, AS CR. He specializes in electron microscopy.

Dr R Cerny is an Associate Professor at the Department of Structural Mechanics, FCE CTU Prague. He works in the field of development of measuring methods for determination of thermal and hygric properties of building materials.

INTRODUCTION

Water vapour permeability δ and liquid moisture diffusivity κ are two main hygric parameters of concrete. They are measured in laboratory conditions mostly, and on the samples which are not exposed to any mechanical load. This is in a clear contradiction to the real situation in a building where unloaded concrete structures appear only exceptionally.

In this paper, δ and κ are determined both with unloaded samples and samples mechanically loaded up to 90% of compressive strength. The water vapour permeability is determined on the basis of the measured amount of water entering or leaving the sample during specified time intervals. The transient method of data analysis utilizing the solution of inverse problem of water vapour diffusion in a material derived recently in our group is employed. The liquid moisture diffusivity is measured using a simple method developed recently in our laboratory which consists in assuming piecewise constant moisture diffusivity with respect to the moisture content, and solving an inverse problem of moisture conduction.

METHODS FOR MEASURING THE HYGRIC PARAMETERS

Moisture Diffusivity

For determination of moisture diffusivity κ we employed a simple method based on the assumption that κ can be considered as piecewise constant with respect to the moisture density ρm (PCK method in what follows). Contrary to the most frequently used methods for for κ determination (see [1 - 4], the PCK method is very fast even for materials with low κ and in addition it exhibits a reasonable precision [5]. Therefore, its application for cement mortar is very suitable. The basic idea of the method is as follows.

We assume that the geometry of the experiment is so designed that the problem of moisture transport can be reduced to only one dimension. Thus, we can write in certain (not very wide) range of moisture the transport equation in the form

$$\frac{\partial \rho m}{\partial t} = \kappa \frac{\partial^2 \rho m}{\partial x^2} \tag{1}$$

The initial and boundary conditions can be defined as

$$\rho m(x,0) = \rho^2 \tag{2}$$

$$\rho m(0,t) = \rho^1 \tag{3}$$

$$\rho m(d,t) = \rho^2 \tag{4}$$

Where ρ_2 is the initial moisture density in the specimen, ρ_1 is the maximum moisture density which can be achieved in the material (the bottom surface of the board is in direct contact with water during the moistening process).

The diffusion problem (1) - (4) can be solved analytically with the result (see [6]):

$$\rho_m(x,t) = \rho_1 + (\rho_2 - \rho_1)\frac{x}{d} + \frac{2}{\pi}\sum_{n=1}^{\infty}\frac{\rho_2 - \rho_1}{n}\sin\left(\frac{n\pi x}{d}\right)\exp\left(-\frac{\kappa n^2 \pi^2 t}{d^2}\right). \quad (5)$$

The total mass of water which penetrated into the sample during the time interval [O, r] can be expressed by the relation

$$m_m(\tau) = S \cdot \int_0^d (\rho_m(x,\tau) - \rho_2)\,\mathrm{d}x. \quad (6)$$

After substituting (5) into (6) we get the final transcendent equation for κ

$$m_m(\tau) - S(\rho_1 - \rho_2)\frac{d}{2} + \frac{2dS}{\pi^2}(\rho_1 - \rho_2)\sum_{n=1}^{\infty}\frac{1}{n^2}(1 - \cos(n\pi)) \cdot \exp\left(-\frac{\kappa n^2 \pi^2 \tau}{d^2}\right) = 0, \quad (7)$$

Which can be solved by some of the iterative methods, such as the Newton method. The value of κ determined by the solution of (7) we award to a characteristic average value of the moisture density in the time interval (O, r),

$$\overline{\rho_{m,c}} = \frac{m_m(\tau)}{2Sd} + \frac{\rho_2 + \rho_1}{2}. \quad (8)$$

In practical measurements we perform the experiment with a set of samples with various values of the initial moisture density ρ_2, and determine the corresponding set of values of the moisture diffusivity κ ($\rho_{m,c}$). In this was we obtain a pointwise given κ (ρ_m) function, i.e., the dependence of the moisture diffusivity on the moisture density.

Water Vapour Permeability

The measuring method for determination of the water vapour permeability δ is, in principle, similar as that for κ (see [7] for details). The measuring apparatus consists of two airtight glass chambers separated by a plate-type specimen of the measured material (see Figure 1). In the first chamber, a state near to 100% relative humidity is kept (achieved with the help of a cup of water), while in the second one there is a state close to 0% relative humidity (set up using some absorption material, such as silica gel). The changes in the mass of water in the cup, Δm_w, and of the silica gel, Δm_a, are measured in dependence on time. In the case that also steady-state measurements are required, the validity of the condition $|\Delta m_w| = |\Delta m_a|$ is tested and the experiment continues until this condition is realised. The experiment is carried out under isothermal conditions. Compared to the classical cup method, employed in most of the European and American standards, our experimental setup has the advantage that it is not necessary to keep the constant relative humidity in whole the climatising chamber but only in a relatively small chamber, and that also the flux of water vapour incoming the specimen is measured. As a consequence, even the steady-state measurements in our setup can be significantly faster than in the classical cup-method set up.

In the practical determination of δ, we know both incoming and outgoing fluxes of water vapour as functions of time, and assume that δ is constant. As follows from the described experimental setup, also the boundary conditions are of the same type as in the case of κ measuring, the constant initial conditions can be managed easily, for instance using dry samples. Therefore, we have mathematically the same problem as in (1 - 4) and the solution is the same as for κ. The final transcendent equation for δ reads.

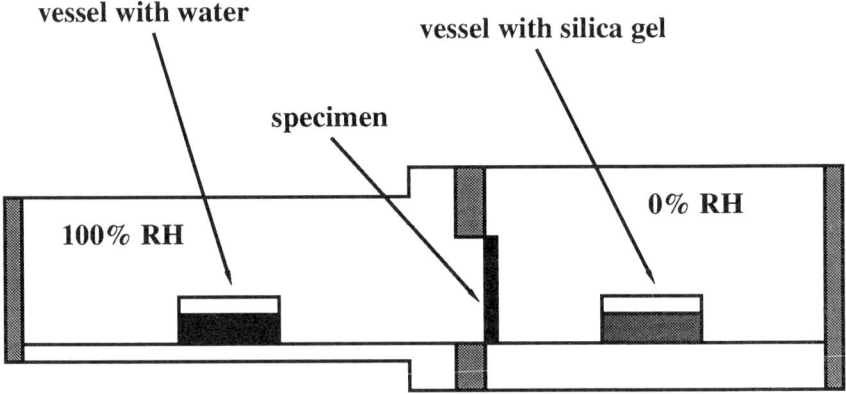

Figure 1 Apparatus for measuring the water vapour permeability

$$m_v(\tau) - \frac{M}{RT}\frac{2Sd}{\pi^2}\sum_{n=1}^{\infty}\frac{1}{n^2}(p_s - p_c\cos(n\pi))\cos(n\pi)\left(1 - \exp\left(-\frac{\delta RT n^2\pi^2\tau}{Md^2}\right)\right) +$$

$$+\frac{M}{RT}\frac{2S}{\pi}\sum_{n=1}^{\infty}\frac{1}{n}\left(1 - \exp\left(-\frac{\delta RT n^2\pi^2\tau}{Md^2}\right)\right)\cos(n\pi)\int_0^d p_o(x')\sin\left(\frac{n\pi x'}{d}\right)dx' -$$

$$-\frac{M}{RT}(p_s - p_c)\frac{S\delta\tau}{d} = 0, \qquad (9)$$

where $m_v(r)$ is the mass of water vapour which was absorbed by the dessicant during the time interval r, R is the universal gas constant, T is the temperature in kelvin, M the molar mass of water, $p_s(T)$ is the saturated pressure of water vapour at temperature T, p_o is the initial partial pressure of water vapour in the sample, p_c is the partial pressure of water vapour in the chamber with dessicant (usually $p_c = 0$), d is the thickness of the sample in the direction of water vapour flow.

MATERIAL SAMPLES

In the experimental measurements of δ and κ, we studied the samples of cement mortar which was chosen instead of real concrete mainly for its better homogeneity, considering the dimensions of samples necessary for the hygric experiments. The composition of the mixture for one charge was the following: Portland cement ENV 197-1, CEM I 42.5 R (Kraluv Dvur, CZ) - 450 g, natural quartz sand with continuous granulometry I, II, III (the total screen residue on 1.6 mm 2%, on 1.0 mm 35%, on 0.50 mm 66%, on 0.16 mm 85%, on 0.08 mm 99.3%) - 1350 g, water - 225 g.

The mortar was prepared by mixing and compacting using mixing machine and vibrator. The samples had cylindrical shape with the diameter of 105 mm and the height of 20 mm. They were left in moulds for the first 24 hours in a high relative humidity environment under wetted cloth. After mould removal, the time remaining to 28 days spent the samples in 20°C water and then they were put in protected external environment (a metal sheet shed) with the relative humidity approximately 65%. After 28 days, the compressive strength was determined on selected samples (57.4 MPa). One part of the remaining samples was mechanically loaded, the other samples were left without any load.

The practical measurements of each of hygric parameters were performed on 20 samples which were approximately 8 months old. One fourth of the samples was not exposed to any load (we will denote it NL in what follows), the second fourth was mechanically loaded to 38.7% of compressive strength transversely (i.e., the load in the direction parallel to the base of the cylinder - it will be denoted as T38), the third fourth to 79.1% of compressive strength transversely again (it will be denoted as T79), and the final fourth to 90.2% of compressive strength standing (i.e., the load in the direction of the normal vector of the base of the cylinder - it will be denoted as S90).

EXPERIMENTAL RESULTS

The experimental results of measurements of hygric parameters are summarized in Table 1. Apparently, both δ and κ were almost unaffected by the mechanical load in a relatively wide range; for the loads up to approximately 80% of compressive strength the differences were within the errorbar of experimental measurements. First the load of 90% of compressive strength has led to significant changes in the hygric parameters, more pronounced was this effect for liquid moisture diffusivity which increased by one order of magnitude compared to the lower values of load.

In order to analyse the reasons for these remarkable changes in hygric parameters, we used the electron microscopy. Scanning electron microscope Jeol JXA-733 was employed to study the structural changes in the surface region of the samples induced by the mechanical load. Typical results are as shown in Figures 2 - 4. Figure 2 represents the sample without any load in the 1000 enlargement. Apparently, no visible cracks appear on the picture. Figures 3 - 4 show the sample loaded in the standing position to 90% of compressive strength (S90) in two basic resolutions corresponding to 1000 and 4800 enlargement, respectively. The typical crack shown in these figures has the width of approximately 1 - 2 µm. The appearance of such wide cracks has to affect the hygric parameters in a high extent, and it is quite logical that their influence on liquid moisture diffusivity which is usually very low in concrete and cement mortar is more pronounced than on the parameters of much faster vapour transport.

Table 1 Hygric parameters of cement mortar

SAMPLE	κ (10^{-9}m^{2}s^{-1})	δ (10^{-12}s)
NL	3.1	3.25
T38	4.7	3.09
T79	4.2	3.15
S90	49.0	3.84

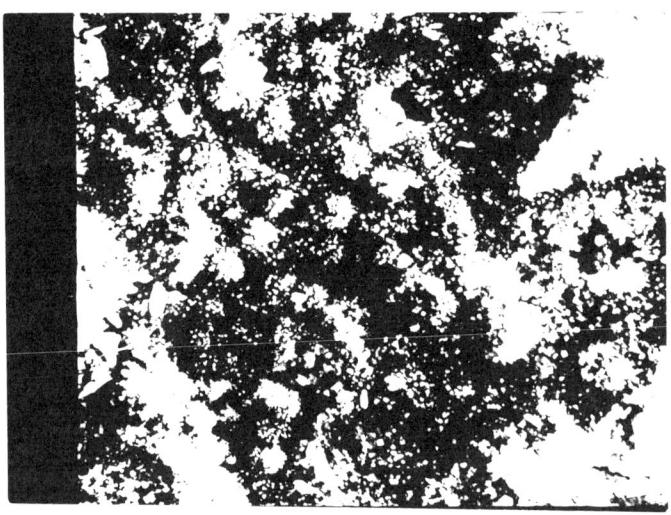

Figure 2 SEM image of an unloaded sample - 9 mm represents 10 µm on the sample

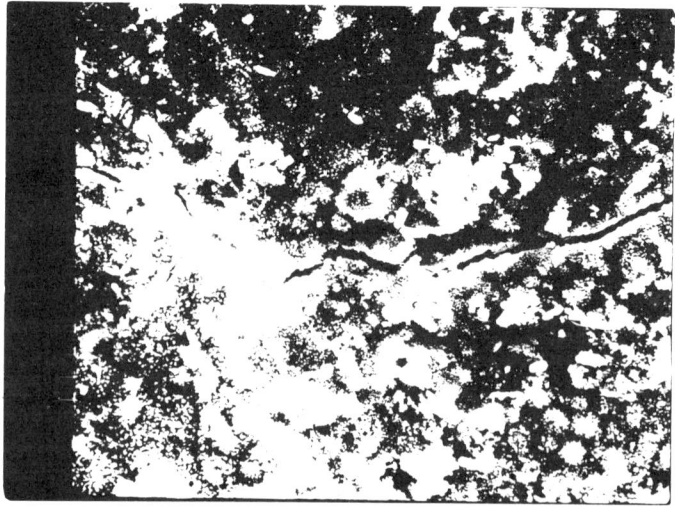

Figure 3 SEM image of a sample loaded to 90% of compressive strength (S90) - 9 mm represents 10 µm on the sample

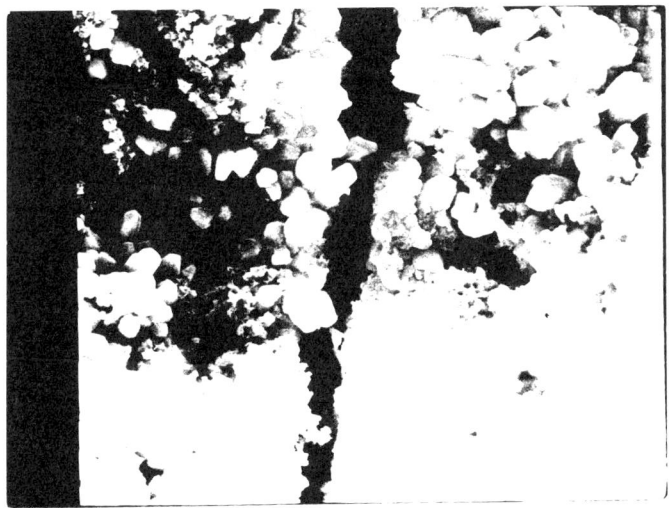

Figure 4 SEM image of a sample loaded to 90% of compressive strength (S90) – a detail, 4.4 cm represents 10 µm on the sample

CONCLUSIONS

The dependence of two main hygric parameters of concrete, namely the water vapour permeability δ and liquid moisture diffusivity κ was determined in a wide range of mechanical load which was first applied on the sample and then released. It has been observed that up to 80% of compressive strength, no significant changes in either κ and δ appeared. Significant effects were first noted at the level of 90% of compressive strength.

From the practical point of view, this appears to be a very good result since few concrete structures are loaded to such a high extent. However, it should be noted that the experimental results presented in this paper concentrated on immediate load application only, and the measurements were performed after load removal. Further experimental work is necessary to analyse the influence of long-term loads, for example, with the measurements of hygric parameters performed during the load exposure.

ACKNOWLEDGEMENTS

This research has been supported by the Grant Agency of the Czech Republic, under grants # 103/97/0094 and # 103/97/K003.

REFERENCES

1. MATANO, C. On the Relation between the Diffusion Coefficient and Concentration of Solid Metals. Jap. J. Phys., Vol. 8, 1933, pp 109.

2. KASPAR, I. Moisture Transport in Building Materials (in Czech). DrSc Thesis, CTU Prague, 1984.

3. HAUPL, R, AND STOPP, H. Ein Beitrag zum Feuchtigkeitstransport in Bauwerksteilen. Schriftenreihe der Sektion Architektur der TU Dresden, Vol. 16, 1980, pp 93.

4. DRCHALOVA, J, AND, CERNY, R. Non-Steady-State Methods for Determining the Moisture Diffusivity of Porous Materials. Int. Comm. Heat and Mass Trans. Vol. 25, 1998, pp 109.

5. DRCHALOVA, J, CERNY, R, AND, HAVRDA, J. Testing of Methods for Determining Moisture Diffusivity of Building Materials. Proc. of Workshop 1998, Part 2, CTU Prague 1998, pp 473.

6. CARSLAW, H, S, AND, JAEGER, J, C. Conduction of Heat in Solids, 2nd Ed., Claxendon Press, Oxford, 1960.

7. CERNY, R, HOSKOVA, S, AND, TOMAN, J, A. Transient Method for Measuring the Water vapour Diffusion in Porous Building Materials. Proc. of International Symposium on Moisture Problems in Building Walls, V P de Freitas, V Abrantes (eds.). Univ. of Porto, 1995, pp 137.

TOUGHENING MECHANISMS AND FRACTURE CHARACTERISTICS OF CONCRETE

V Bílek

ÚVAR Brno Joint Stack Company

Z Keršner

Technical University of Brno

Czech Republic

ABSTRACT. This paper deals with the toughening mechanisms of fracture of cement-based composites. The crack deflection, the distributed interfacial cracking, the crack bridging and trapping are considered. The mathematical models by Lange-Kornbak and Karihaloo are used. Test on concrete specimens made from Portland cement and aggregates of three fractions were arranged. Specimens were cured in different state conditions during the year. The distinction between the contribution of toughening mechanisms of paste - mortar - concrete is discussed with respect to microstructure. The scanning electron microscopy (SEM) and roentgen diffraction analysis (RDA) were used. The uncertainties in material, geometrical and loading conditions of tested specimens are taken into account. Input variables of mathematical models are considered as random ones. The random variables are described by the type of probability distribution function, mean value and standard deviation. Statistical analyses of the toughening models are performed using the Monte Carlo type simulation technique based on the Latin Hypercube Sampling (LHS) method.

Keywords: Cement-based composite, Fracture characteristics, Effective fracture toughness, Toughening mechanism, Microstructure, Random variable, Simulation.

Dr Vlastimil Bílek is a Research Fellow in Building Materials Laboratory, ÚVAR Brno, Joint-stack Company, Czech Republic. He specialises in the field of technology of cement-based composites, fracture properties and microstructure.

Dr Zbyněk Keršner is a Lecturer at the Institute of Structural Mechanics, Faculty of Civil Engineering, Technical University of Brno, Czech Republic. His main research interests include the numerical analysis of structures with respect to concrete technology, the fracture characteristics of concrete.

INTRODUCTION

As is well known, cement paste shows nearly brittle fracture, while concrete shows significant non-linear behaviour. That is, some toughening mechanisms are active. Four toughening mechanisms of the fractured cement-based composites have been described in the literature. These mechanisms are discussed below. The properties of concrete develop with age and according to curing. In the opinion of the authors, changes in this development play an important role in the fracture of concrete. The specification of toughening mechanisms can help to explain some interesting changes. It is possible to estimate which is the weak link of cementitious composite – the paste or the interface between paste and aggregates, or the weak aggregates.

Important changes of concrete characteristics were recorded during first year of curing. This paper is focused on the specification of toughening mechanisms and the explanation of these phenomena.

MATHEMATICAL MODELS OF TOUGHENING MECHANISMS

The influence of various toughening mechanisms on the fracture characteristics of cement-based composites is studied. The mathematical models by Lange-Kornbak [1] based on Li [2] are used. These models can work with the concrete/matrix effective fracture toughness ratio (r_{theor}). The effective crack model of Karihaloo and Nallathambi (see e.g. [3]) was used for the determination of effective fracture toughness K_{Ic}^{e}. Thus it is possible to distinguish four toughening mechanisms controlling tension softening: The distributed interfacial cracking (microcracking; with notation M_1), the crack deflection (M_2), the crack bridging (M_3) and the trapping (M_4). In a symbolic way we can write

$$\frac{K_{Ic,concrete}^{e}}{K_{Ic,matrix}^{e}} = \sqrt{M_1.M_2.(M_3+M_4)}$$

$$M_1 = f(V_{agg}, v_{matrix}, v_{concrete})$$

$$M_2 = f(V_{agg})$$

$$M_3 = f(V_{agg}, g_{av}, f_{t,agg}, v_{matrix}, v_{concrete}, K_{Ic,matrix}^{e})$$

$$M_4 = f(V_{agg})$$

where V_{agg} is the total volume fraction of aggregate (the unit aggregate content to density of aggregate ratio m_{agg}/ρ_{agg}), g_{av} is the average aggregate size, $f_{t,agg}$ is the uniaxial tensile strength of aggregate and v is Poisson's ratio. Input variables of these models are summarised in Table 1.

The theoretically obtained values of the ratio r_{theor} were compared with the measured value r_{test} obtained from the test for concrete (or mortar) and paste specimens. This procedure was repeated for each age of concrete. This enables the distinction of different toughening mechanisms controlling the fracture at different ages. The authors have applied this procedure to various types of cements [4].

STATISTICAL APPROACH

The uncertainties in material, geometrical and loading conditions of tested specimens are taken into account. Input variables of the mentioned mathematical models are considered as random ones. Statistical analyses of the toughening models were performed using the Monte Carlo type simulation technique based on the Latin Hypercube Sampling method: See [5] and [6] for details. All random input variables of toughening models and the particular set of their statistical parameters (mean value, standard deviation, probability distribution function (PDF)) are presented in Table 1, where Φ is the uncertainty factor of the model.

Table 1 Random input variables of toughening models

VARIABLE	UNIT	MEAN VALUE	STAND. DEVIATION	PDF
$K^e_{Ic,paste}$	MPa.m$^{1/2}$	0.6	0.07	normal
V_{paste}	-	0.22	0.01	log-normal
$V_{concrete}$	-	0.22	0.01	log-normal
$f_{t,agg}$	MPa	10	1	normal
m_{agg}	kg	1924	96.2	normal
ρ_{agg}	kg/m^3	2600	130	normal
g_{av}	mm	8.9	0.45	normal
Φ	-	1	0.05	normal

MATERIALS, MIX PROPORTIONS AND CURING ENVIRONMENTS

Portland cement (CEM I 42.5 R) was used in our study. One type of high strength concrete was used. The water/cement ratio, $w/c = 0.32$ and cement content was 400 kg/m^3. A super-plasticizer was also used. Three fractions of aggregates were used: Sand of 0-4 mm and crushed gravel of 8-16 mm and 11-22 mm.

Mortar was made according to the above mix proportion without crushed gravel and the w/c ratio reduced to 0.30 because water wetting of aggregates was taken into account. For the same reason, the w/c ratio for paste was only 0.24. Mortar and paste were cured in fog room.

Four series of test specimens were prepared by the above mentioned composition, which differed in curing conditions:

Series K - steam curing, maximum temperature $t_{max} = 60°C$ and curing in outdoor environment,

Series E - steam curing, $t_{max} = 75°C$ and curing in outdoor environment,

Series N - normal curing in outdoor environment,

Series L - normal curing in laboratory environment (fog room).

For fracture tests of concrete and mortar beams 80 x 80 x 480 mm (depth x width x length) were made. The width of beams from paste was 60 mm only. The effective fracture toughness K_{Ic}^e was determined for ages of specimens 1, 7, 28, 90 and 365 days. Because the load-deflection diagram was continuously recorded, the fracture energy G_F can be computed from the area below this curve. A method for including the tail of the curve was used [7].

It is evident that the dimensions of the specimen are smaller than those recommended. Only the comparison of the measured values is significant for the purposes of this paper. Deviations of the measured values are discussed below.

RESULTS AND DISCUSSION

The values of effective fracture toughness for four different series are presented in Figure 1. A more conspicuous change is visible in the development of fracture energy - see Figure 2. It is evident that K_{Ic}^e and G_F show an interesting development, and it is necessary to present the variation of toughening mechanisms.

The concrete/matrix effective fracture toughness ratios r_{theor} were computed in accordance with the toughening models. Fracture toughness of the matrix (paste or mortar) and composite (mortar or concrete) was also measured experimentally. We suppose, there are active the mechanisms of toughening (or its combination) during failure of concrete with best accord between experimentally and theoretically obtained values of r_{test} respectively r_{theor}. From Table 2 it can be seen that combinations of toughening mechanisms differ for different ages and various specimens. The notation, e.g. M_{123}, in this table indicates of combination of three toughening mechanisms in fracture: Microcracking (superscript 1), crack deflection (superscript 2) and crack bridging (superscript 3).

The development of fracture toughness of paste is interesting. In order to explain it, the roentgen diffraction analysis (RDA) and scanning electron microscopy (SEM) were used. The back-scattered electron imaging was very useful. The method of crack quantification was developed by Bílek and Janová [8]. In this method, the content of microcracks is expressed by the ratio: crack area to global area of micrograph. The microcrack content increases from 7 to 365 days as can be seen in Figure 3. We suppose that the autogenous shrinkage controlled the increase in microcracks and this attacked the fracture energy and other characteristics.

Differences between toughening mechanisms for mixes K, L and N at the same age are not significant. More significant are the differences of toughening mechanisms contribution during ageing of concrete. It is evident, more of the mechanisms are active in age 7 or 28 days. After, a decrease of number of active mechanisms occur. Between 28 and 365 days the fracture toughness of paste and mortar increase, but fracture toughness of concrete decrease. What is reason of this development of fracture toughness?

We believe, a delayed hydration of cement grains in hardened paste initiate microcracks development [9], namely between 28 to 90 days. After this period, the healing of the microcracks occur in cement paste by new product of delayed cement hydration. The hydration is source of autogenous shrinkage of paste.

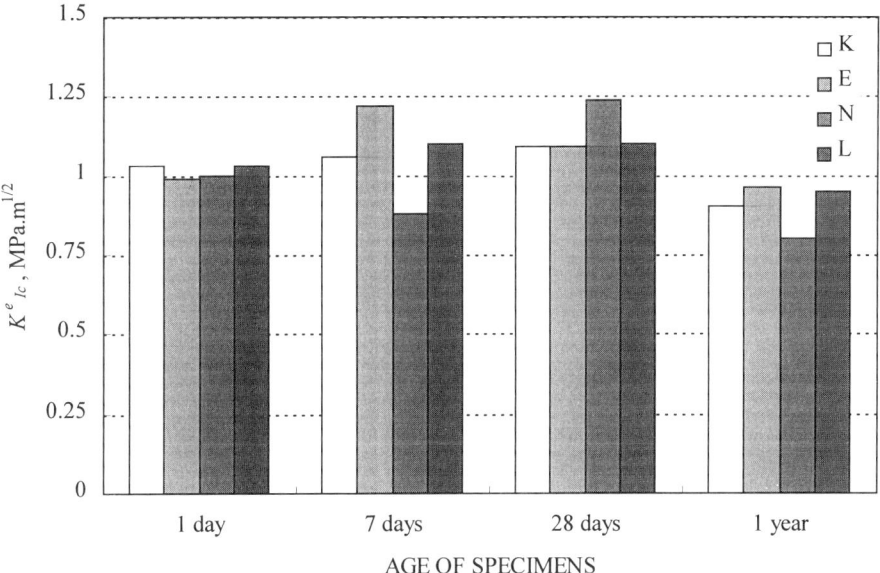

Figure 1 Fracture toughness with age for different series of specimens

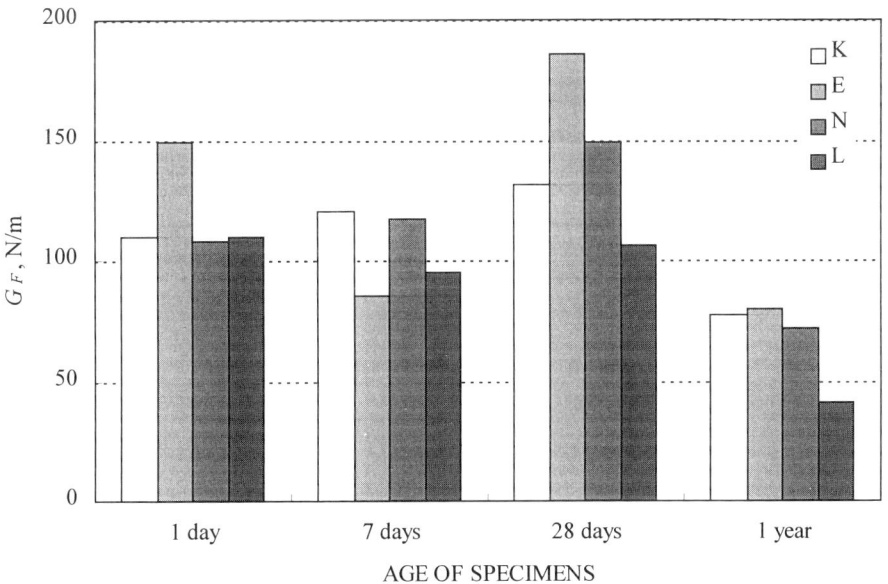

Figure 2 Fracture energy with age for different series of specimens

Table 2 Effective fracture toughness of matrix (mean value (standard deviation)) and toughening mechanisms

			7 days	28 days	90 days	1 year
			AGE OF SPECIMENS			
K	$K_{Ic\ paste}^{e}$ MPa.m$^{1/2}$		0.61 (0.03)	0.60 (0.02)	-	-
	Specimen	1	M_{13}	M_{123}	-	-
	(concrete/paste)	2	M_{123}	M_{34} or M_{123}	-	-
		3	M_{123}	M_{12}	-	-
L	$K_{Ic\ paste}^{e}$ MPa.m$^{1/2}$		0.61 (0.03)	0.60 (0.02)	0.72 (0.09)	-
	Specimen	1	M_{34}	M_{12}	M_{2}	-
	(concrete/paste)	2	M_{12}	M_{34}	M_{23}	-
		3	M_{23}	M_{12}	M_{2} or M_{1}	-
N	$K_{Ic\ paste}^{e}$ MPa.m$^{1/2}$		0.28 (0.03)	0.45 (0.02)	0.26 (0.03)	0.56 (0.05)
	Specimen	1	M_{1234}	M_{234}	-	M_{13}
	(concrete/paste)	2	M_{1234}	M_{234}	-	M_{123}
		3	M_{1234}	M_{234}	-	M_{123}
N	$K_{Ic\ mortar}^{e}$ MPa.m$^{1/2}$		0.6 (0.06)	0.71 (0.03)	0.65 (0.01)	1.24 (0.03)
	Specimen	1	M_{34}	M_{123}	-	$r_{test} = 0.73$
	(concrete/mortar)	2	M_{34}	M_{23}	-	$r_{test} = 0.78$
		3	M_{123}	M_{23}	-	$r_{test} = 0.81$
N	$K_{Ic\ paste}^{e}$ MPa.m$^{1/2}$		0.28 (0.03)	0.45 (0.02)	0.26 (0.03)	0.56 (0.05)
	Specimen	1	M_{123}	M_{2}	M_{34}	M_{123} or M_{234}
	(mortar /paste)	2	M_{123}	M_{3}	M_{234}	M_{234}
		3	M_{12}	M_{2}	M_{234}	-

Difference between paste shrinkage and constant volume of the aggregates caused a new microcracks rise in the concrete. This script of degradation of fracture characteristics is more detailed studied in [10].

What about mechanisms of toughening? See mixture N. At the age of 7 days a number of toughening mechanisms is active between concrete and pastes. The number decreases gradually. The same tendency we can see in the case of toughening between concrete and mortar. A difference from this tendency is visible in the case of toughening between mortar and paste. Now age of 1 year.

When we compare the mechanisms of toughening between paste and concrete and mechanisms of toughening between mortar and paste, we can see all toughening of cement composites is controlled by cement paste. At age of 1 year the fracture toughness of concrete is lower than fracture toughness of mortar and r_{test} is less than one.

Figure 3 Microcracks quantification: Area of cracks vs. age of specimens for selected series

CONCLUSIONS

1. Cement hydration controls not only strength development. Its influence on fracture characteristics is evident.

2. Decrease of fracture toughness (and fracture energy too) is connected with microcracks formation.

3. In our opinion, the microcracks content bring a change of toughening mechanisms.

4. Combinations of more toughening mechanisms are active at early ages.

5. On the base of difference between fracture toughness development of matrix (pastes, mortar) and composite (mortar, concrete) a possible explanation of degradation of fracture toughness is proposed.

It is probable that concrete durability is strongly controlled by the above mentioned phenomena. This is the reason why the autogenous shrinkage effect must be limited. The method of cracks quantification and toughening mechanisms specification are suitable for studying this effect.

ACKNOWLEDGEMENTS

The financial support under the Technical University of Brno Project No. FV 280004/98 and the Grant Agency of the Czech Republic Project No. 103/97/K003 are gratefully acknowledged. The authors would like to express their sincere thanks to Mr. T. Mosler and Mr. J. Dufek from the Building Materials Laboratory (ÚVAR Brno) for their careful preparation of the tested specimens.

REFERENCES

1. LANGE-KORNBAK, D. AND KARIHALOO, B. L. Design of concrete mixes for minimum brittleness. Advanced Cement Based Materials, No. 3, 1996, pp. 124-132.

2. LI, V. C. AND HUANG, J. Relation of concrete fracture toughness to its internal structure. Engineering Fracture Mechanics, Vol. 35, No. 1/2/3, 1990, pp. 39-46.

3. KARIHALOO, B. L. Fracture mechanics of concrete. Longman Scientific & Technical, New York, 1995.

4. KERŠNER, Z. AND BÍLEK, V. Influence of microstructure on toughening mechanisms of concretes. Engineering Mechanics, Vol. 5, No. 3., 1998, pp. 199-201.

5. BÍLEK, V., KERŠNER, Z. AND SCHMID, P. Fracture mechanics of aggregate-paste interface. Proc. Int. Symp. Brittle Matrix Composites BMC5, A. M. Brandt, V. C. Li, I. H. Marshall eds., Warsaw, 1997, pp. 300-309.

6. NOVÁK, D., TEPLÝ, B. AND KERŠNER, Z. The role of Latin Hypercube Sampling method in reliability assessment of structures. Proc. Int. Conf. on Structural Safety and Reliability ICOSSAR'97, Kyoto, 1997, 403-406.

7. ELICES, M., GUINEA, G. V. AND PLANAS, J. On the measurement of concrete fracture energy using three-point bend tests. Materials and Structures, Vol. 30, 1997, pp. 375-376.

8. BÍLEK, V. AND JANOVÁ, D. Quantification of microcracs by means of electron microscopy. Proc. of Conf. CONCON'98, Vol. 2, Prague, 1998, pp. 97-101 (in Czech).

9. IGARASHI, S. AND KAWAMURA, M. Reduction in strength in high strength mortars at long ages. Proc. of Conf. FRAMCOS-3, Vol. 1, Gifu, 1998, pp. 243-252.

10. BÍLEK, V. Possibility of explanation of interesting development of fracture properties of concrete. Proc. of Conf. Life Prediction and Ageing Management of Concrete Structures, Bratislava, 1999, in print.

STUDY ON INFLUENCES OF SURFACE MICORCRACKS ON PERMEABILITY AND FROST RESISTANCE OF STEAM CURED CONCRETE PRODUCTS

M Aba   M Shoya
Hachinohe Institute of Technology
K Otsuka
Tohoku Gakuin University
Japan

ABSTRACT. The purpose of this study is to investigate the influences of surface microcracks on permeability and frost resistance of steam cured concrete products. First, the characteristics of microcracks, which were formed during the steam curing period and enlarged during the subsequent drying period, were examined by a new X-ray technique using a contrast medium and SEM. Then, the influence of microcracks on steam cured specimens were investigated on the water permeability, the air permeability and the freezing and thawing resistance. From this investigation, it was clarified that the microcracks formed on the surface of steam cured concrete could be classified into different three types, that is, separation cracks between aggregate and paste, mortar cracks and vapor cracks. Then, it was confirmed that the microcracks affected the rate of transport of water and gasses into concrete, and that the frost resistance of concrete was rapidly lowered when the crack density represented by the ratio of microcracks exceeded a certain threshold value.

Keywords: Microcracks, Steam cured concrete products, Permeability, Frost resistance, X-ray inspection technique.

Dr Minoru Aba is an assistant professor in the Department of Civil Engineering at Hachinohe Institute of Technology, Aomori, Japan. He received his Dr. Eng. from Tohoku Gakuin University in 1997 and has been engaged in research work on microcracks formed on concrete.

Dr Masami Shoya is a professor in the Department of Civil Engineering at Hachinohe Institute of Technology, Aomori, Japan. He received his Dr. Eng. from Hokkaido University in 1984 and has written many papers on concrete durability and shrinkage.

Dr Koji Otsuka is a professor in the Department of Civil Engineering at Tohoku Gakuin University, Sendai, Japan. He received his Dr. Eng. from Tohoku University in 1981 and has authored many papers on the properties of bond and cracks of reinforced concrete members.

INTRODUCTION

In recent years, the precast concrete structural members have come to be adopted extensively due to rationalization of construction, elimination of labor and so on. Generally, the precast concrete structural member is produced with atmospheric steam curing. However, microcracks are sometimes formed on the surface of concrete during the steam curing period, which can be hardly detected by the naked eye. Then, if concrete products suffer early drying, the microcracks may grow larger in their dimensions. This will result in a lower durability of the precast concrete structure. In recent years, the actual cases for the early deterioration of concrete products have been reported. One of the reason will be related to the microcracks formed due to the initial stress developed during the steam curing procedure. However, there exists no proper technique to detect such microcracks, and characteristics of the microcracks and the relationship between the formation of microcrack and the durability of concrete have not been fully investigated.

The objective of this study is to investigate the influences of surface microcracks on durability of steam cured concrete products. First, the characteristics of microcracks, which were formed during the steam curing period and enlarged during the subsequent drying period. A new X-ray technique using a contrast medium was developed for this study to detect microcracks [1]. Then, relationships between surface microcracks and durability of concrete were studied by water permeability test, air permeability test and freezing and thawing resistance test.

MATERIALS USED IN TESTS AND TESTING METHOD

Concrete Materials and Specimen

The cement used was high-early-strength portland cement whose specific gravity and specific surface area by Blaine were 3.13 and 424 m^2/kg, respectively. The fine aggregate was a river sand whose specific gravity, water absorption and fineness modulus were 2.54, 2.60 % and 3.01, respectively. The coarse aggregate was crushed stone whose specific gravity, water absorption and fineness modulus were 2.69, 2.01 % and 6.57, respectively.

The mixture proportions were given by a water-cement ratio of 0.5, with two types of air content of 2 percent and 5 percent as shown in Table 1. The dimensions of specimen used were 100 by 100 by 400-mm prisms as shown in Figure 1.

Table 1 Mix proportions of concrete

TYPE	MAX AGG SIZE (mm)	SLUMP (mm)	AIR CONT. (%)	W/C	s/a (%)	UNIT CONTENT (kg/m³)				AE
						Water	Cement	Fine Agg	Coarse Agg	
Non AE	20	80±10	2	0.50	49	205	410	805	885	-
AE	20	80±10	5	0.50	46	192	384	744	923	0.096

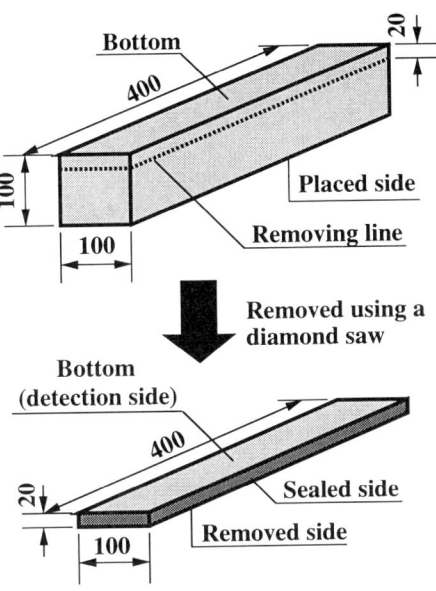

Table 2 Steam curing conditions

	PRE-STEAM PERIOD (hr)	RATE OF TEMP RISE (°C/hr)	PEAK TEMP AT PERIOD		COOLING RATE (°C/hr)
			(°C)	(hr)	
1	0	12	55	5.8	-3
2	2	12	55	4.9	-3
3	4	12	55	4.0	-3
4	4	12	55	4.0	-3
5	4	20	55	4.9	-85
6	4	35	55	5.5	-3
7	6	12	55	3.1	-3
8	Ordinary Moist Curing (OMC)				

Figure 1 Dimensions of specimen (mm)

Curing Conditions

Table 2 shows the steam curing conditions used in the tests. Four types of different presteaming period, three types of rate of temperature rise and two types of rate of cooling were adopted. The other conditions such as, the presteaming temperature and the maximum temperature were constant at 20°C and 55°C. Besides, the ordinary moist curing condition in water at 20°C to the age of 7 days, was used for a comparison. Then, secondary curing was made from the end of steam curing and ordinary moist curing. Two curing conditions were adopted for secondary curing, one is drying in 20°C, 50 percent relative humidity, and another is submerging in water in 20°C untill the required age.

Detection of Microcracks

X-ray technique using a contrast medium

After the curing period, the bottom 20-mm of the specimen were removed using a diamond saw (Figure 1), and the surface microcracks formed during the curing period were detected. A X-ray technique using a contrast medium was used to detect the surface microcracks. The contrast medium was permeated into the microcracks formed on the surface of the concrete. The X-ray inspection was made, and microcracks were detected by a sharkasten (film viewing device) from X-ray shadowgraph film [1], [2]. The total length of the cracks was measured by the curvimeter on traced cracks, and the density of microcrack length was calculated as the total length of cracks divided by the traced area (100 by 100-mm).

Then, as shown in Figure 1, all the surfaces of the specimen, except the bottom (detection side of the microcracks) were sealed with epoxy resins. The specimens were placed in the controlled room for 91 days to replicate the secondary curing period. During the secondary curing period, X-ray inspection was made at 3, 7, 14, 21, 28, 56 and 91 days after secondary curing.

SEM

The surface microcracks were also observed by SEM at the end of the steam curing period and at the end of the subsequent drying period of 14 days, 28 days. The maximum width of surface microcracks measured on SEM photographs and their distribution was investigated. Then, from these results, average and standard deviation of the surface microcrack width were calculated, and the density of microcrack area was defined as the average of surface microcrack width multiplied by the density of microcrack length. The SEM used is Wet-SEM in which the surface processing procedure of the sample is unnecessary.

Water permeability Test

Water permeability was measured by the in-put method proposed by Murata [3]. The experiment was started at the end of the steam curing period and at the end of the subsequent drying period of 14 days, and was made applying a water pressure of 980 kPa to an area of concrete having a 100 by 100-mm (detection side of the microcracks) for 48 hours using 100 by 100 by 100-mm specimens cut from 100 by 100 by 400-mm prisms. After the completion of the test, the specimens were split, and the average depth of the penetration of water was measured. Then, the diffusion coefficient was calculated by the theoretical equation proposed by Murata.

Air Permeability Test

Air permeability was measured by the constant pressure method proposed by Nagataki and Ujike [4]. The experiment was started at the end of the steam curing period and at the end of the subsequent drying period of 14 days, and was made holding a load bearing pressure of 490 kPa to an area of concrete having a 100 by 100-mm (detection side of the microcracks) using 100 by 100 by 100-mm specimens cut from 100 by 100 by 400-mm prisms. The amount of air penetrating the concrete specimens was captured using the water replacement method. The coefficient of air permeability was calculated by the theoretical equation applying Darcy's Law.

Freezing and Thawing Test

Freezing and thawing test was conducted in accordance with ASTM C 666-92 procedure A (freezing and thawing in water) [5]. The test started at the end of the steam curing period and at the end of the subsequent drying period of 14 days, 28 days using 100 by 100 by 400-mm prisms. Three specimens were tested simultaneously according to the same testing conditions. However, all the surfaces of the specimen except the bottom were sealed with epoxy resins in order to allow deterioration only from the bottom side (detection side of the microcracks). The change in the relative dynamic modulus of elasticity was measured every 30 cycles up to the 300th cycle. The dried specimen was submerged in water for 1 week after the drying period and then sealed before the test was started. From these examination results, durability factor DF was calculated.

EXPERIMENTAL RESULTS AND CONSIDERATIONS

Characteristics of Microcracks

An example of the cracks traced from a X-ray film (100 by 100-mm), which was obtained by the X-ray technique using a contrast medium at the end of the steam curing period, was shown in Figure 2. It can be seen from this figure that at the end of the steam curing period there existed many microcracks on the surface of the concrete. On closer investigation, surface microcracks could be classified into three types, separation cracks between aggregate and paste, mortar cracks and vapor cracks [1]. In the following, enlarged shadowgraphs of these cracks shown on the X-ray film are presented.

The shadowgraph of separation cracks between aggregate and paste was shown in Figure 3. In the photograph, the arcshaped white line around the aggregate is the separation crack where the contrast medium permeated into chinks between the aggregate and the paste. The shadowgraph of mortar cracks was shown in Figure 4. The gray part in the center is an aggregate, the white curve around the aggregate is the separation crack, and the cloudy parts around the separated aggregate are considered as groups of many microcracks formed in the mortar. In Figure 5, the spherical part in the center is an air void, and the radial white lines starting from the air void are the vapor cracks.

Influence of Steam Curing Conditions on Microcrack Formation

The results of measurement of microcracks were presented in Table 3. The density of microcrack length tends to increase with the decrease of presteaming period, and with the increase of rate of temperature rise and cooling, at the end of the steam curing period. When the presteaming period of 0 hour, relatively many mortar cracks in addition to separation cracks were detected, and a few vapor cracks were also detected. In the case of the presteaming period of 6 hours, almost all of the cracks detected were separation cracks. Then, these microcracks were observed by SEM, when the at the end of the steam curing period, the averages of maximum width of surface microcracks were almost less than 10-μm.

The changes in the density of surface microcrack length during the subsequent drying period was shown in Figure 6. From this figure, the densities of surface microcrack length rapidly increased during the first month of the drying period after steam curing. After 1 month, the rates of increase slowed down. In the case of the microcracks formed during the drying period, both separation cracks and mortar cracks were detected.

These microcracks were observed by SEM, and as shown in Table 3, the widths during subsequent drying after steam curing were distributed with large width and wide deviation than those at the end of the steam curing period. However, when the specimens were submerged in water after the steam curing period, the microcracks gradually were decreased with the increase of submerged period.

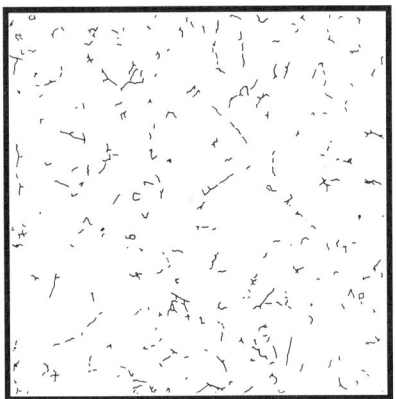

Figure 2 Microcracks traced from X-ray film

Figure 3 Separation cracks between aggregate and paste

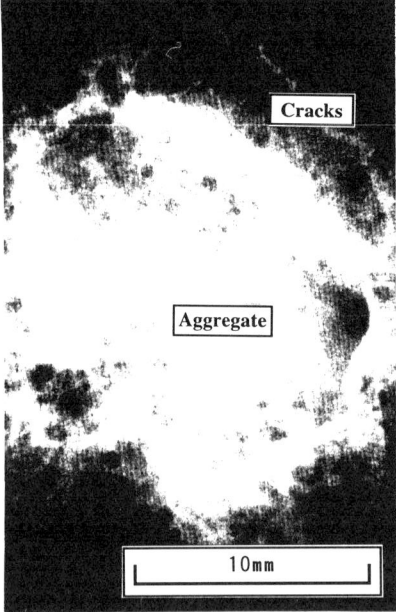

Figure 4 Mortar cracks

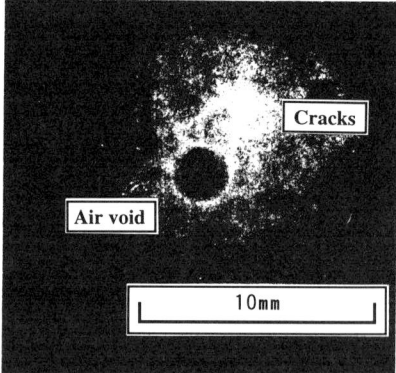

Figure 5 Vapor cracks

Water Permeability

The relationship between diffusion coefficient and average of microcrack width, which was plotted with regard to the steam curing condition at a fixed air content of 5 %, is shown in Figure 7. At the end of the steam curing period, the averages of microcrack width were grouped around 0.008-mm, and the diffusion coefficients were almost the same value. Also the diffusion coefficient was rapidly increased when the microcrack width exceeded approximately 0.06-mm in case of the drying period of 28 days after the steam curing period.

Table 3 Results of measurement of microcracks

	PRE-STEAM (hr)	RATE OF TEMP RISE (°C/hr)	RATE OF COOLING	TYPE	AT END OF STEAM CURE PERIOD			AT END OF 28 DAY DRYING PERIOD		
					Density (mm/cm²)	Width Mean (μm)	Width S.D. (μm)	Density (mm/cm²)	Width Mean (μm)	Width S.D. (μm)
1	0	12	-3	NonAE	7.50	10.9	9.9	18.50	80.1	10.7
				AE	7.30	9.7	6.8	19.10	77.9	16.8
2	4	12	-3	NonAE	1.65	6.9	8.5	8.89	69.2	13.3
				AE	1.67	6.6	4.7	8.94	62.2	13.3
3	4	35	-3	NonAE	2.98	8.3	5.8	11.18	74.2	13.6
				AE	3.04	7.0	4.9	11.36	70.4	14.9
4	4	12	-85	NonAE	4.78	7.4	4.2	14.33	75.4	13.9
				AE	4.74	7.9	5.6	14.26	73.2	15.8
8	Ordianry moist curing			NonAE	1.01	5.1	3.6	5.37	66.3	12.4
				AE	0.91	5.4	4.6	5.52	65.9	14.3

S.D = standard deviation

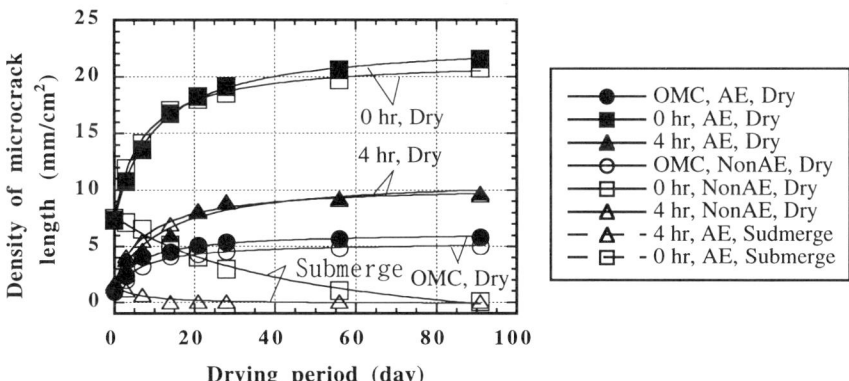

Figure 6 Change in density of surface microcrack length during dying period after steam cure

The relationship between diffusion coefficient and density of microcrack length was shown in Figure 8. When the microcrack width is very small at the end of the steam curing, the diffusion coefficient is not almost changed with the increase of density of microcrack length. However, When the microcrack width was enlarged subjected to drying for 28 days after the steam curing period, the diffusion coefficient was increased with the increase of density of microcrack length from the neighborhood of 5-mm/cm².

This results will give that a certain threshold value is existed of the microcrack width and the density of microcrack length which the influence on water permeability becomes very little.

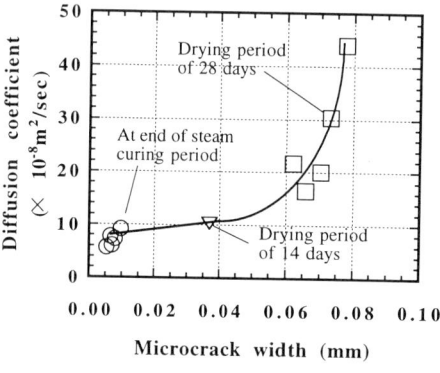
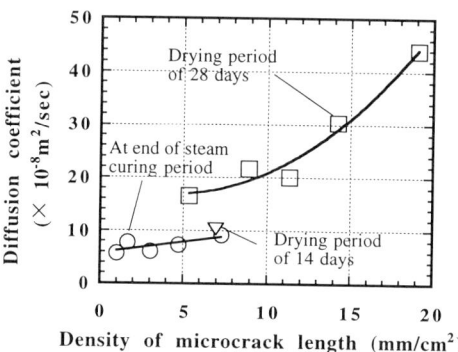

Figure 7 Relationship between diffusion coefficient and average microcrack width

Figure 8 Relationship between diffusion coefficient and density of microcrack width

Air Permeability

The relationships between coefficient of air permeability and microcrack formation at a fixed air content of 5 % were shown in Figure 9, Figure 10 and Figure 11. The coefficient of air permeability was increased with the increase of microcrack width and density of microcrack length, in case both at the end of the steam curing period and at the end of the subsequent drying period of 28 days. And, it can be seen that the coefficient of air permeability was increased in the fairly good correlation with the increase of the density of microcrack area.

From those results, it was confirmed that the rate of transport of gasses into concrete is not relatively dependent on the width of cracks. This will be probably caused by that the gasses is permeated into relatively narrower gap of concrete than that in case of water.

Freezing and Thawing Resistance

The relationship between durability factor DF and average of microcrack width was shown in Figure 12. At the end of the steam curing period, the averages of microcrack width were grouped around 0.008-mm, and the durability factors were almost the same value. On the other side, durability factors were rapidly decreased when the microcrack width exceeded approximately 0.06-mm in case of drying period of 28 days after the steam curing period.

The relationship between the durability factor DF and density of microcrack length is shown in Figure 13. It can be seen that when the microcrack width is very small at the end of the steam curing period, the durability factors are little changed with the increase of density of microcrack length. However, when the microcracks width enlarged from 28 day drying, the durability factors of both AE concrete and Non-AE concrete were decreased with the increase of density of microcrack length from the neighborhood of 5-mm/cm^2.

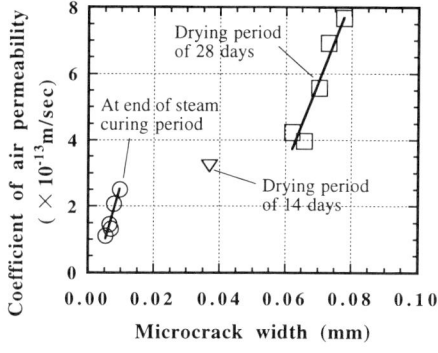

Figure 9 Relationship between coefficient of air permeability and average microcrack width

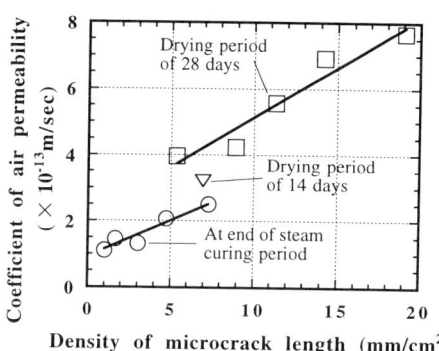

Figure 10 Relationship between coefficient of air permeability and density of microcrack length

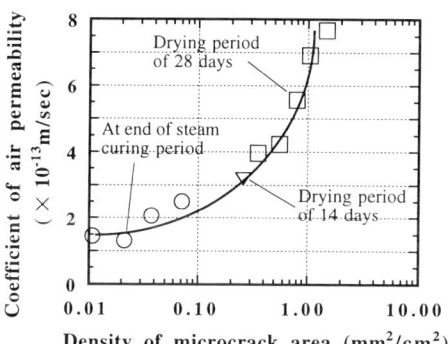

Figure 11 Relationship between coefficient of air permeability and density of microcrack length

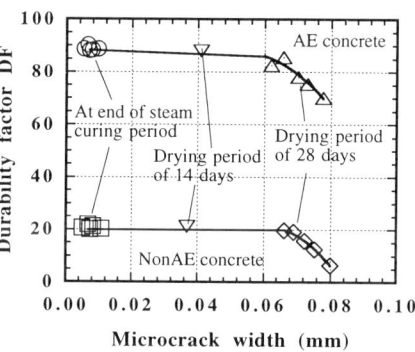

Figure 12 Relationship between durability factor DF and average microcrack length

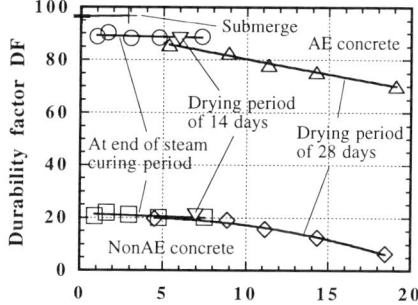

Figure 13 Relationship between durability factor DF and density of microcrack length

The reason is guessed that the quantity of freezing water was little increased by the fact that the water could be hardly permeated into the concrete when the width of microcracks was very small. This results will indicate that a certain threshold value is existed of the microcrack width and the density of microcrack length which the influence on frost resistance became very lower.

CONCLUSIONS

Experiments were carried out by X-ray technique using a contrast medium and SEM to detect the surface microcracks formed during the steam curing period and enlarged during the subsequent drying period. Relationships between surface microcracks and durability of concrete were also studied. The following conclusions can be drawn.

1. Using a X-ray technique with contrast medium, the surface microcracks formed on the steam cured concrete could be detected. The surface microcracks could be classified into three types, such as separation cracks between aggregate and paste, mortar cracks and vapor cracks.

2. The shorter presteaming period is likely to form more microcracks. In case of the shorter presteaming period, relatively many mortar cracks in addition to separation cracks were detected. The density of surface microcrack length rapidly increased during the 1 month of the drying period after steam curing. After the 1 month, the rates of increase slowed down.

3. It was confirmed that the microcracks affected the rate of transport of water and gasses into concrete. The rate of transport of gasses into concrete is relatively dependent on the width of cracks different from the case of water penetration. This will be probably due to the fact that the gasses is easily permeated into relatively narrow gap when compared with the water.

4. The frost resistance of concrete was rapidly lowered when the crack density of microcrack exceeded a certain threshold value.

REFERENCES

1. OTSUKA, K., SHOYA, M., KOSEKI, K. and ABA, M., Properties of microcracks formed on surface of concrete at steam curing period. Journal of Materials, Concrete Structures and Pavements of JSCE, No.520, V-28, 1995, pp.143-155 (In Japanese).

2. OTSUKA, K., SHOYA, M. and ABA, M., Influences of surface microcracks on durability of steam cured concrete. Journal of Materials, Concrete Structures and Pavements of JSCE, No.585, V-38, 1998, pp.97-111 (In Japanese).

3. MURATA, J., Studies of the Permeability of Concrete. Proceedings of Japan Society of Civil Engineers, No.77, 1961, pp.69-103 (In Japanese).

4. NAGATAKI, S. AND UJIKE, I., Air-Permeability of Concrete, Cement & Concrete, No.455, 1985, pp.24-31 (In Japanese).

5. ASTM, Standard Test Method for Resistance of Concrete to Rapid Freezing and Thawing (ASTM C 666-92), Annual Book of ASTM Standards, Section 4 Construction, Volume 04.02 Concrete and aggregates, 1996, pp.224-236.

INVESTIGATION OF TRANSITION ZONES BETWEEN ORDINARY CONCRETE SURFACE AND DIFFERENT REPAIR MATERIALS

K Pettersson
Cement and Concrete Research Institute
Sweden

ABSTRACT. One of the most common repair technique for chloride induced corrosion reinforcement is to remove the chloride contaminated concrete just around the corroded area on the steel and rebuild it with some repair mortar. Theoretically, when the rebar is in contact with two different concrete qualities, the difference in potential will rise and it is easier for the corrosion to start and create new corrosion damage. This type of corrosion damages is called macro cell corrosion. Such damages is not so common in the real concrete structures.

This paper presents results from an investigation where the focus is on the transition zone between repair mortar and ordinary chloride contaminated concrete. The repair mortars are cement based with polymeric components, inhibitors, and different water cement ratio. Chloride penetration and corrosion potential measurement and alkali diffusion are going to be measured in this studied. The investigation started in the winter 1997.

Keywords: Concrete repair, Transition zones, Durability, Corrosion

BSc Karin Pettersson is project leader and researcher at the Swedish Cement and Concrete Research Institute in Stockholm, Sweden. She specialises on the chloride induced corrosion and also assessment of residual service life of reinforced concrete structures

INTRODUCTION

Deterioration of concrete structures is a problem of major concern in countries where the infrastructure follows the coast-line or is subjected to the action of deicing salts. The associated problems of deterioration have led to an increasing need for maintenance and repair of these structures. Experience has shown that chloride induced corrosion of reinforcement is the dominant cause of damage [1]. The most widespread repair technique for chloride contaminated concrete in Sweden and most other countries is to remove covercrete with an unacceptable chloride content and rebuild it by applying a repair mortar.

A lot of research on the repair of reinforced concrete has been concerned with alternative materials to replace contaminated and spalled concrete and to improve the properties of these materials from an isolated point of wiev, without taking into account the influence of steel condition, substrate condition, properties of the transition zones between the materials and the steelmortar interface. The exception from this is the dimensional behaviour of repair materials relative to the substrate.

Vassie [2] found that the durability of conventional repairs are very sensitive to steel condition, i.e. the steel cleaning procedure. If chloride contaminated corrosion products are not removed during the repair operation it is possible that the reinforcement will continue to corrode. In this investigation no effective methods for preventing further corrosion on steel was used. Only bruching was used on the rebar before the repair materials were used.

The importance of the steel-mortar interface has been reviewed by Vaysburd [3]. The protection of reinforcement is depending on the buffering action of a lime-rich layer at the steel-mortar interface and the bond between steel and concrete or repair mortar. A dense steel-mortar interface with good bond will block the accumulation of corrosion products, the OH^--ions are easily supplied to the steel surface by the calcium hydroxide at the interface, and the movement of Cl^--ions needed to sustain pitting is limited. When voids are present at the interface, all protective mechanisms provided by the concrete or mortar are weakened and corrosion is promoted.

Discussions about the conventional repair technique, in which cover concrete is removed and the profile is restored only at the severely deteriorated parts of the structure, may result in accelerated rebar corrosion due to macrocell corrosion. Macrocell corrosion may develop either as a consequence of insufficient removal of chloride contaminated or carbonated concrete, or by penetration of aggressive substances into the weakest parts of the repair. The first mentioned possibility has been demonstrated by Schiessl and Breit [4] in laboratory experiments. The two other possibility for macrocell corrosion to occur have not been subject for many research investigations, but from a theoretical and logical point of view the probability for such problems to occur should be significant.

The present paper discusses results from an experimental work, focusing on transition zones between chloride contaminated concrete surfaces and repair mortars, and with special attention to the edges between repaired and non-repaired area. Seventy two concrete slabs were rebuilt with eleven different repair materials. The repair materials were chosen because of it's diffusivity (open or closed) and also because of the inhibiting effect on the rebar corrosion.

EXPERIMENTAL DETAILS

Preparation and Repair Work

The concrete slabs, about 500 x 150 x 150 mm, were six years old. They have been submerged in artifical 3% chloride seawater in the laboratory. The water cement ratio, w/c, vary from 0.40 to 0.75. Some of them have additives as micro silica, 5-10% and fly ash, 25% of the cement weight. Before the repair was carried out an area of 30 or 60 cm^2 of half the slabs was removed to a specified depth of 10 mm under the rebar, for repair as shown in Figure 1.

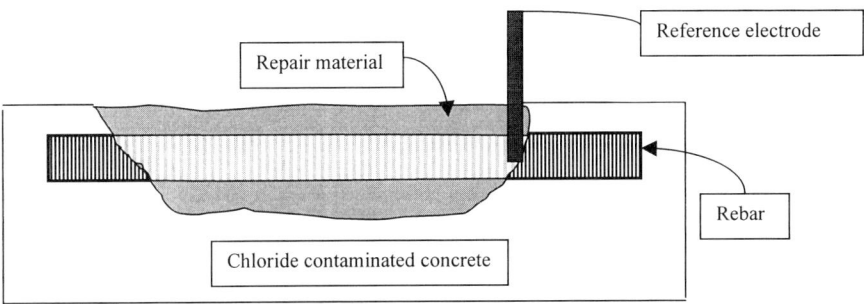

Figure 1 Repaired concrete slab

After the concrete removal was completed, the surfaces were cleaned from loose particles and evaluated with respect to roughness. The surfaces were found to be very suitable for further repair procedures.

An referenselectrode was cast in 20 of the repaired slabs. It was a mangan oxide electrode that is commonly used in concrete. The reference electrode was placed just near the transition zone, Figure 1. Eleven commercial repair systems were selected, of which five were dry-shot mortars and six were surface treatment. The repair materials were applied by experienced personnel. During mixing, placing and curing, the instructions and recommendations from the materials producers were followed exactly in order to avoid defects caused by bad workmanship. The repaired slabs were cured in 100% RH and covered with plastic for one month before they were stored in a climate room with 80% RH.

Materials

The five repair mortars used in this investigation were all commercial ready-mixed products, i.e. only water had to be added. Some mortars were cement based and contained conventional additives, such as silica fume, water-reducing agents, inhibitors and superplastisizers. The aggregates were well-graded with a maximum size of 8 mm.

Exposure Conditions and Testing

After the slabs were cured, half of the slabs were submerged in artifical 3% chloride seawater and the other half were exposed for salt spray. The specimens were water saturated before exposure. The corrosion potential is measured continously with the reference electrode and a logger system. This method is often used to investigate whether the embedded rebar is corroding or not. The corrosion potential gives a qualitative assessment about the corrosion presence. The measured corrosion potential values can often be difficult to interpret since the concrete micro climate has a major influence on the measurements. The measured values are generally interpreted in accordance with the American standard ASTM G876, which defines the likelihood of corrosion as shown in Table 1.

Table 1 Likelihood of corrosion at various measured potentials

POTENTIAL, mV (Cu/CuSO4)	CORROSION RISK
- 200 and more positive	10%
- 200 to –350	uncertain
- 350 and more negative	90%

The most important is to measure the potential continously and observe when a big change will appear, either in positive or negative direction.

EXPERIMENTAL RESULTS

Corrosion Potentials

The results demonstrate the ability of three different repairsystem and the w/c ratio to influence the corrosion. After i month in 100% RH and 40 day in 80% RH the slabs were exposed to artifical seawater.

In Figure 2 it can be seen that the concrete slabs with the lowest w/c ratio have the highest potential. It can also be seen that the drop in potential is biggest for the slabs with w/c higher than 0.4.

When using inhibitor in the repairsystem the potential will increase with time after the slabs have been exposed to the seawater.

In Figure 3 no significant differeces between the repair areas could be seen. Index B in the legend means a big repair area about 60 cm^2. Index S means a small repair area of about 30 cm^2.

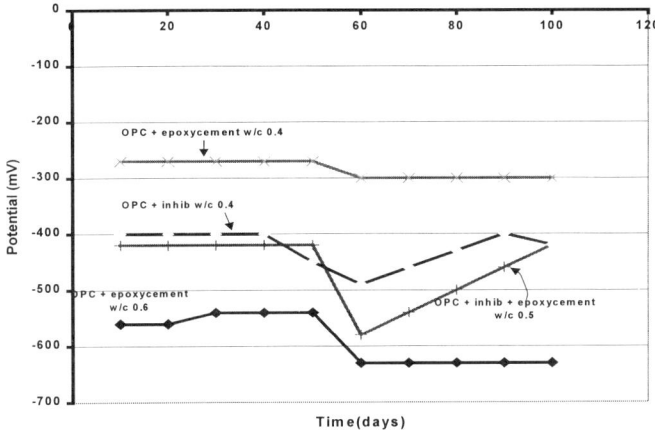

Figure 2 Corrosion potential versus the time, w/c and different repair system

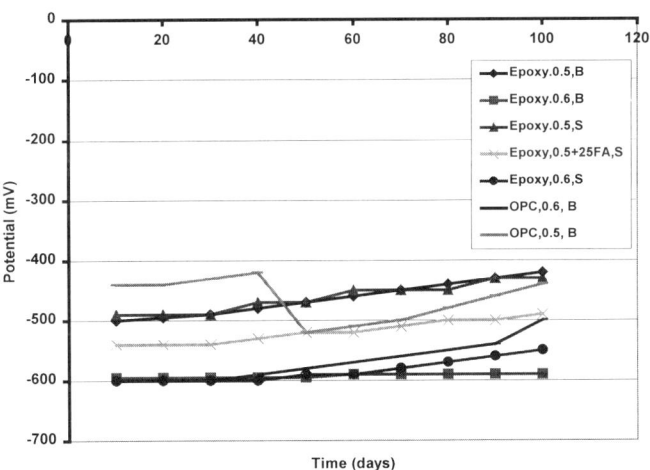

Figure 3 Corrosion potential versus the time and size of repair area

As can be seen in Figure 4, where only surface treatements have been used, the potential will be more and more positive with time. It´s difficult to explain this but some of the treatments will diffus into the concrete during time and this will influence the micro climate in the concrete and also the corrosion potential in a possitive way.

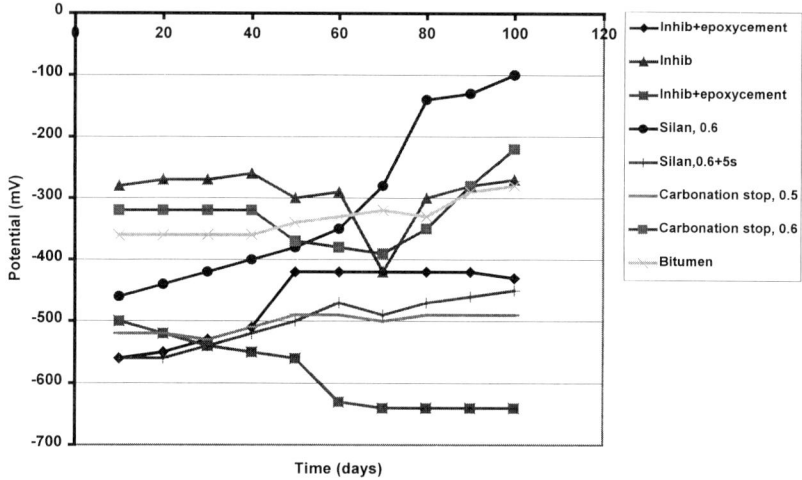

Figure 4 Corrosion potentials versus the time and different surface treatments

SUMMARY AND DISCUSSIONS

The present paper reports results from an experimental work, focusing on transition zones between chloride contaminated concrete, repair mortars and surface treatments.

The corrosion process can be stoped by reducing the water-, chloride- and oxygen content into the concrete. This can be made by the above mentioned repair systems. In this investigation it can be seen that different repair systems will act different compare to the corrosion potential. Some of them will be very effective due to corrosion activity. About the transition zone, no data have been found yet. From repaired field concrete structures we have found that the transition zon is not as distinct as we thought and that the macro cell effect will not occur as in the theory. One explanation for this can be the diffusion of hydoxyl ions from the repair mortar to the chloride contaminated concrete. This will result in a more soft transition zone between the concrete and the repair material. This maybe can reduce the macro cell effect for corrosion. Opposite results have been found by Pedersen [5]. Generally, using hand-applied mortars, the results show significant improvements when using a cementious bonding agent.

The surface treatment seems to make the corrosion potential more positive which also results in less corrosion activity. Due to the difficulties of measuring exactly in the transition zones, the quantitative information from the investigation is limited but the results should be reliable and clear. When this paper is written the exposure time is not more than one year. This is probably to short for evaluate any significant effect on the different repair system. The investigation will go on for more than two years from now and than more results will be reported.

REFERENCES

1. PETTERSSON, K Service life of concrete structures – in a chloride environment. CBI-report 1:97, 1997.

2. VASSIE, P R. The Influence of Steel Copiditioll on The Effectiveness of Repairs to Reinforced Concrete, Construction & Building Materials Vol.3 No.4, December 1989, pp 201-207.

3. VAYSBURD, A M. SomeDurabilityConsideratiotisforEvaluatingandRepairingConcrete Structures, Concrete International, March 1993, pp 29-35.

4. SCHIESSL, P, BREIT, W. Local Repair Measiires at Concrete Structures Damaged by Reinforcement Corrosion - Aspects of Durability, Fourth International Symposium on Corrosion of Reinforcement in Concrete Construction, Robinson College, Cambridge, England, 1996, pp 525-534.

5. PEDERSEN, V. Investigation on transition zones between hydrodemolished concrete surfaces and different concrete repair materials. International Conference in Svolvaer, Norway 28-30 May 1997. pp 329-338.

THEME TWO:
ACHIEVING PERFORMANCE

Keynote Paper

INTERFACES IN CONCRETE - ACHIEVING PERFORMANCE

S Subramanian
Larsen & Toubro Limited
India

ABSTRACT. Frontiers of concrete performance have been expanding in the last few decades. Concrete of higher strength and better durability are being proportioned, manufactured and used. The parameters of concrete performance can be altered for the better by choosing proper materials or by modifying the physical interfaces between the materials.

The utility of concrete can be further enhanced by interfacing the interested segments of society: viz., scientist and engineer, specifier and supplier and client and contractor.

This paper gives a survey of the role of physical interfaces and outlines a few concepts regarding societal interfaces, for improving concrete performance.

Keywords: Concrete performance, Physical interfaces, Abrasion, Impact, Durability, Strength, Societal interfaces, Scientist, Engineer, Supplier, Specifier.

Mr S Subramanian is Senior Deputy General Manager (Research & Development), Construction Group of Larsen & Toubro Limited, Chennai, India. His present responsibilities include design of concrete chimneys and cooling towers, apart from development work on construction materials and concrete. His research interest is in use of chemical and mineral admixtures in concrete.

INTRODUCTION

In the last five decades the engineering profession and the public have witnessed ever increasing levels of performance from concrete as well as concrete specialist. In tune with other technical developments there has been a shift towards mechanisation and centralised production. If concrete structures with high standards are to be built, we may consider performance at two types of interfaces viz.,

a)   Physical or Technical Interfaces

b)   Societal and Professional Interfaces

We can enumerate and deal with the physical interfaces in comparatively straight forward way. These physical interfaces are detailed in Figure 1.

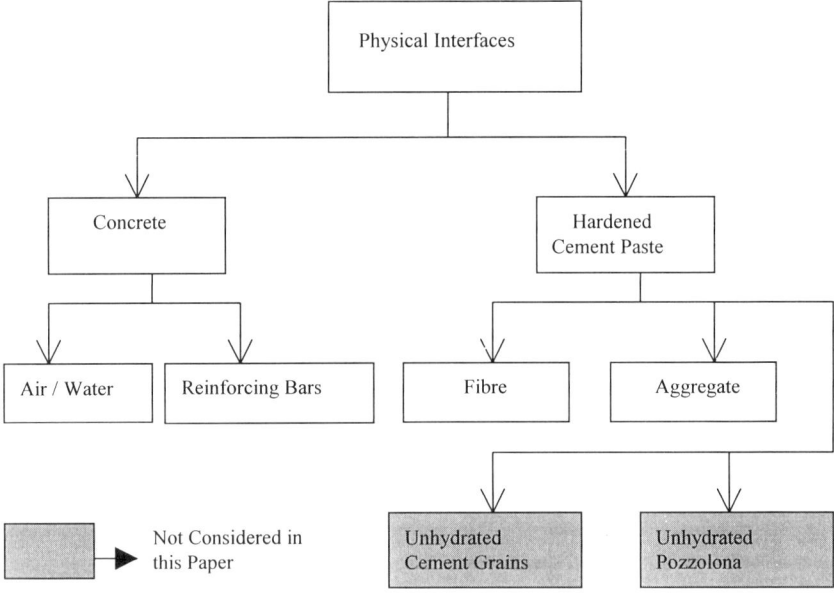

Figure 1 Physical interfaces

The social and professional interfaces that affect the performance of a concrete structure are complex. We have quite a number of trades and interests exerting conflicting pushes and pulls. The owner, user, contractor, architect, specification writer, design engineer and the scientist interact among themselves and with other groups.

Their interaction, communication and commitment to a common goal will affect the societal or professional interfaces and the overall result (Figure 2).

PHYSICAL INTERFACES

Beginning with the researches of Farras in France in the early 1950s, the transition zone and the aggregate-hardened cement paste (hcp) interface have been studied both experimentally and by computer simulation. The present view is that in a 25 – 50 μm thick zone surrounding the aggregate, a duplex layer of Ca (OH)$_2$ and a porous shell, having properties that are different from the bulk cement paste, are present. This transition zone at the aggregate - matrix interface has the following characteristics (Figure 3).

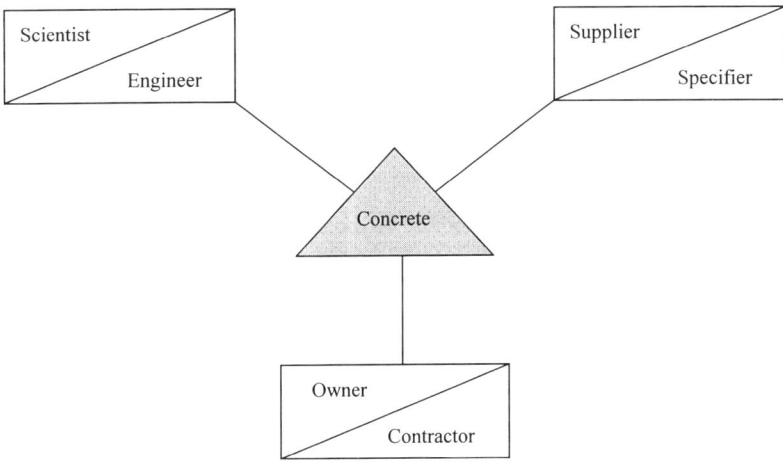

Figure 2 Societal interfaces

(a) In concrete containing excess water, the bleed water tends to collect below the aggregates, thereby creating an unfavourable w/c ratio, a porous region and weak zone (Figure 3a).

(b) Near the large surface of an aggregate, the cement particles are not packed in an ideal manner, thereby leading to more porosity than in the bulk paste (wall effect) (Figure 3b).

(c) As the aggregate surface is inert, the C-H-S gel (formed on the cement grain) grows only in one direction and not towards the aggregate faces (one sided growth) (Figure 3c).

(d) To varying degrees, the factors mentioned above, result in a zone that is considerably more porous than the bulk paste. This encourages the deposition of oriented crystals of calcium hydroxide, giving rise to weak planes.

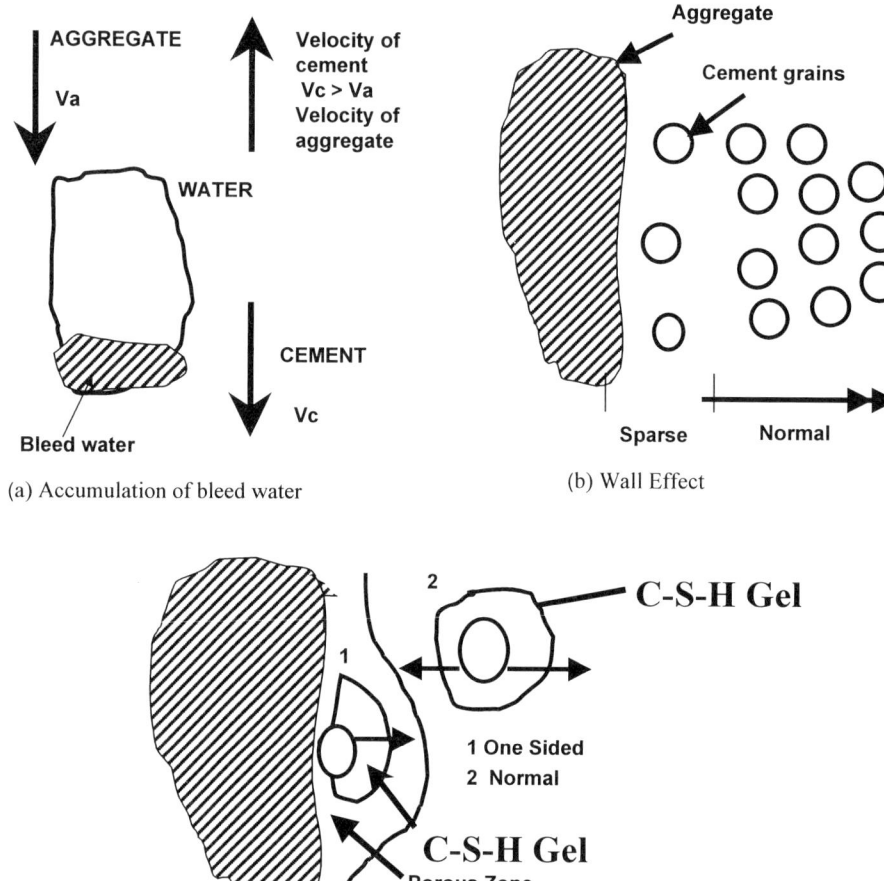

(a) Accumulation of bleed water

(b) Wall Effect

(c) One Sided Growth of C-S-H Gel

Figure 3 Factors leading to porous aggregate-hcp interface

Improving the Aggregate / hcp Interface

Removal of excess water

The effect of bleed water has been well recognised. For example, ACI code, 318-1977, recommends that the development length should be increased for 'top bars' [1]. One more example; the 'optimum' value for maximum size of aggregate is recommended as 10 to 12.5 mm (large aggregates tend to collect bleed water and very small aggregates increase the water demand – 10 mm aggregates seem to strike a balance) [2]. Thus, one can expect to get improved properties of the aggregate / hcp interface, simply by reducing the w/c ratio.

This is confirmed by the results of experiments by Mitsui et al. [3].

When water-cement ratio is reduced from 0.65 to 0.35, the push-out load (a measure of bond strength) was increased from 1200 N to 1600 N, and the porosity, (at a distance of 50 μm) was reduced from 24% to 18%.

A further reduction in water content near the interface can be effected by using aggregates having slightly porous surfaces. The excess water is drawn into these pores and cement particles are also pulled towards surface. This is known as filter effect [4]. Another suggestion is to reduce the surface tension of water through the use of a water reducing agent or surfactant. This will reduce the thickness of the water film on the aggregate.

Improve the packing of small particles near aggregate face

Average size of cement grains can be taken as say 30 to 40 μm. If we add very small particles, say 1 to 5 μm dia, the small particles will fill up the interstices and densify the interface and also improve the bond.

Data from [3] can be cited to demonstrate this. Normal concrete with w/c = 0.35 had a push-out load of 1600 N and a porosity of 18%. Addition of 10% micro silica (w/c = 0.35%), dramatically improved the porosity characteristics to 3% and the push out load to 2600 N.

Coat the aggregate faces with a reactive layer

Zimbelmann [5] suggested that the bond strength of the transition zone might be improved by inducing a chemical or physical reaction between the aggregate and the hcp. This effect is also demonstrated by Mitsui et al. [3], referred earlier. By precoating the aggregate particles with a cement / silica fume slurry, the push-out load was increased from 2600 N to 4200 N and the porosity at the interface transition zone was practically eliminated (Figure 4).

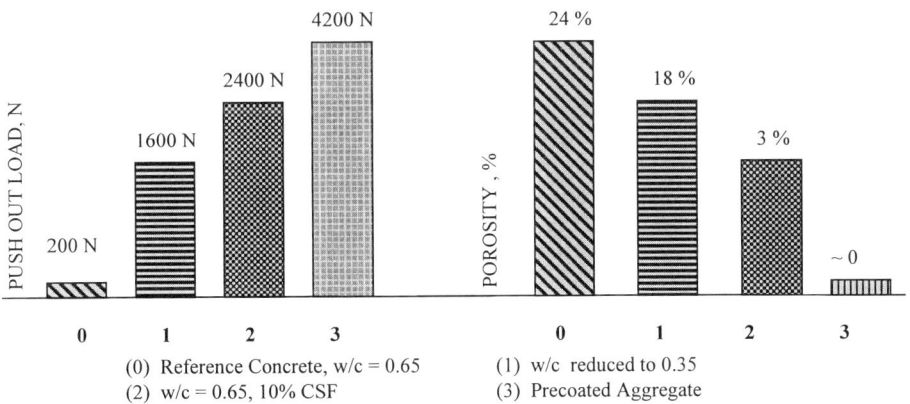

Figure 4 Improvement in porosity and bond strength through selected methods (redrawn from [3])

INTERFACES IN CONCRETE - ACHIEVING PERFORMANCE

The 'Ideal' Aggregate

If one were to impart all the desirable properties to an aggregate and improve the characteristics of the interface between aggregate and hcp, then this 'ideal' aggregate will consist of an inert, hard core, coated with a hydraulic layer that is slightly porous and capable of combining with cement. Shan Yeng et al. [6] prepared such an ideal aggregate consisting of an inert core of corundum, coating it with a mixture of $Ca(CO)_3$ powder, small amount of gypsum and water and firing at $1400°$ C for four hours.

By measuring the electrical conductivity parameter at the interface, they established that the treated corundum had superior properties compared to granite, untreated corundum and Portland clinker.

Improving the Reinforcement – hcp Interface

According to Mindess [7], the steel reinforcement tends to be surrounded by a thin layer of massive $Ca(OH)_2$ crystals. Since adhesion and friction do not contribute much to the bond strength, the details of the microstructure at the reinforcement – hcp interface are of little importance. It is more expedient to develop the required pull-out resistance by increasing either the compressive strength or the embedment length, rather than modifying the interface transition zone.

Improving the Reinforcing Fibre – hcp Interface

Fibres are incorporated in concrete mainly to increase the toughness and not the strength. As in the case of steel reinforcement, there is a thin layer of $Ca(OH)_2$ surrounding the fibres. In the case of fibre reinforcements too, it is not necessary to modify the morphology at the hcp interface.Failure of fibres is generally controlled by pull-out (or bond) strength and not by breaking. Thus it is easier to improve the performance of fibre reinforcement by altering the geometry of the fibres, than by densifying the interface (Figure 5).

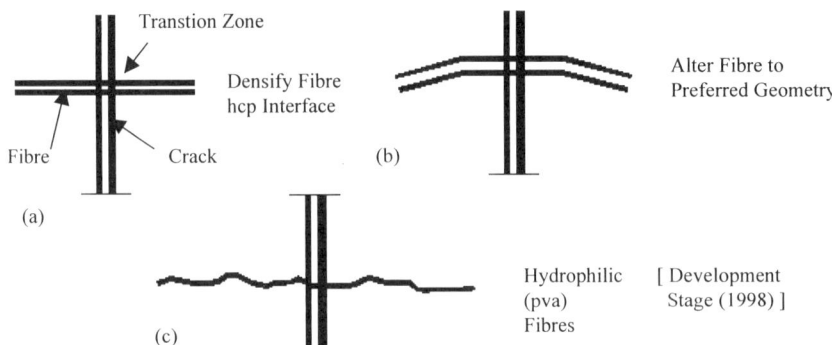

Figure 5 Methods for improving performance of fibres in concrete

A third possibility of improving the performance of fibre reinforced cement composites has also been examined by Kanda & Li [8]. Polyvinyl alcohol (PVA) fibres are hydrophilic and therefore, are able to develop surprisingly high bond strengths through chemical action as well as friction. In this case, the fibres tend to fail by rupture rather than by pull out, because of their high bond strengths and small diameter.

Concrete-Air Interface

This interface deserves close attention, as the resistance of concrete to external agents, both chemical and mechanical, is primarily determined by the quality of concrete that is exposed to the air, water or soil.

Transport properties (permeability and diffusion) of the cover concrete determine the durability of the structure to a large extent, but attention to this topic so far has not been adequate. Concrete near the air interface should ideally be dense, impermeable and non-absorbent. However, accumulated bleed water, laitance and capillary channels render it porous and weak, particularly a zone of about 2mm from the surface. The quality of the 'near surface' concrete determines its abrasion resistance, appearance, resistance to carbonation, transport of oxygen and chloride ions and also the freeze thaw durability.

IMPROVING CONCRETE PERFORMANCE – ROLE OF PHYSICAL INTERFACES

The description of interfaces outlined above, though incomplete, can nevertheless be useful for devising ways to improve the performance of concrete structures.

Appearance

Concrete surfaces that are left without any treatment or coating, tend to accumulate dirt, stain and organic matter, because of the tendency of capillary pores to imbibe and retain water. The remedy is to incorporate a stearate (misleadingly called integral 'water proofer') to provide a water repellant coating at the surfaces of the capillary pores and thus discourage the accumulation of dirt on the concrete surface.

Abrasion and Wear

Three types of abrasion or wear are usually recognised. (a) Mechanical abrasion or gouging, (b) surface wear caused by foot traffic and light pneumatic tyred vehicles, say in warehouses and (c) hydraulic erosion. Regardless of the type of abrasion, compressive strength appears to be the important factor in determining the abrasion resistance of concrete. This can be further improved by the use of hard and dense aggregates both for the upper and lower end of the grading zone [9]. Of course, importance of curing at the early ages cannot be over emphasised.

An interesting possibility of getting enhanced abrasion resistance is by densifying the interface between the hcp and the aggregate and eliminating the weaker $Ca(OH)_2$ crystals.

This can be achieved by using fly-ash or better still, with silica fume. Aluminate cements, having low calcium content, together with hard aggregates like corundum are used for machine parts subjected to high abrasion.

Shock and Impact

Densifying the interface between aggregate and hcp may improve the tensile strength and to some extent the compressive strength. However, when shock and impact loads are expected, we need to have post cracking ductility or toughness rather than strength. Fibres are incorporated in the matrix to improve the toughness and they usually fail by pull-out rather than rupture, as noted earlier.

We come across two approaches to enhance the toughness of the composites. Both attempt to modify the interface.

a) Polymer modification (using a latex) of the matrix to promote better mechanical compatibility by enhancing the bond and increasing the failure strain of the matrix [10].

b) Using hydrophilic fibres, made of polyvinyl alcohol [8].

In the near term, polymer modification of the matrix may emerge as a simpler alternative.

Durability

In respect of durability three factors are to be considered

a) Carbonation of cover concrete

b) Diffusion of chlorides

c) Permeability to liquids

Out of these factors only permeability to liquids may be affected by the cracks in the aggregate matrix interface. When we improve the strength of the interfacial zone by addition of slag / fly ash / pozzolonic materials the pores are refined and thus the permeability – diffusion properties of the matrix are also improved.

Mitsui K. et al. ([3] pp. 119-128) have shown that the porosity at the interface between aggregate and hcp is dramatically reduced from 24% to 3%, by reducing the water cement ratio to 0.35 and incorporating about 10% silica fume.

Bentz D.P et al. ([4] p155) have studied percolation process in the aggregate hcp interface by computer simulation. They observe that modification of the interfacial zone microstructure can alter the percolation characteristics and thereby can improve the durability of concrete with respect to attack by transport controlled process (such as sulfate or chloride ion ingress).

Strength

Performance of concrete in respect of strength parameters can be improved by altering the micro structure of the aggregate hcp interface. Stroeven [11] attributes this improvement to the packing achieved by incorporation of micro fillers, but cautions that better interface bond need not necessarily lead to improve global mechanical properties. Nor will an enhanced compressive strength due to interface modification automatically imply that other mechanical properties will be upgraded. The resulting concrete with higher strength properties, most often will also be brittle.

The inter-relationships between materials, interfaces and concrete performances are summarised in Table 1.

Table 1 Role of physical interfaces in cocnrete performance

	INTERFACES					
	Air	Water	Agg.	Rebar	Fibre	Strategy
1 Appearance	X ✓					Coat capillaries with hydrophilic materials.
2 Abrasion & wear						
a) Gouging			✓ ✓			Increase compressive
b) Wear			✓ ✓			Strength
c) Erosion		X X	✓ ✓			Use pozzolona
3 Shock & impact					✓ X	a) Modify matrix with latex b) Use hydropholic fibre (?)
4 Durability	✓ X	✓ X	✓ X			a) Low w/c ratio b) Effective curing c) Use pozzolona, slag
5 Strength						
a) Bond with rebar				X ✓		a) Increase concrete strength b) Increase embedment
b) Compressive strength			✓ X			a) Low w/c ratio b) Densify Interface c) "Reactive" aggregate with porous surfaces
c) Tensile strength			✓ ✓	X	✓	

Legend

Interface (No Modification)	X	Parent Material (No Modification) X
Interface (Modification required)	✓	Parent Material (Modification Required) ✓

SOCIETAL INTERFACES

Concrete technologists and structural engineers do not exist in isolation from other segments of society and it is not enough that economical solutions on initial cost basis are devised and executed. There is an awareness that our resources, even when abundant are finite and not limitless and therefore, we need sustainable development. Long term concerns dictate that the damage to the environment, due to exploitation of resources, should be minimum.

Our working environment based on 'division of labour' principles has also changed. Today neither the individual brilliance nor the rote effort aligned to assembly line techniques will be adequate. The efforts of individuals are required to be combined into organisational resources to execute the mega projects. For example, the concrete technologist has to interface with the scientists and the users of the technology, namely the consulting engineer, architect and the owner.

Scientist – Engineer Interface

A large number of applied scientists have dealt with various disciplines such as metallurgy, ceramics, silicates and polymers. The influence of the interfaces of concrete technologist with scientists in other discipline can be easily discerned (Table 2).

Table 2   Ideas for developments in concrete

NO	SOURCE OF TECHNOLOGY	RELEVANT AREA	APPLICATION TO CONCRETE
1	Chemical Engineering	Fluidized Bed Reactors	Ideal packing of Aggregates, Cement and Fillers
2	Leather & Textile Technology	Dye Dispersants and Tanning agents	Concrete Admixture (Plasticizers & Superplasticizers)
3	Metallurgy	Powder Metallurgy for tungsten carbide tool tips	Densified Mortar with superior strength and Durability
4	Electro Chemistry	Corrosion Abatement	Cathodic Protection

1. Theory for packing of particles and its influence on their mobility was developed for fluidised beds by chemical engineers. These concepts are today extended to grading of aggregate, cement and microsilica for concrete.

2. Some of the admixtures, mainly plasticizers and super-plasticizers were developed by chemists for leather and textile industries.

3. Densified mortars with superior durability and strength properties have been developed by compressing cement-silica pastes with low w/c rates. Similar techniques are employed by metallurgists also (powder metallurgy).

4. For dealing with the problem of corrosion of reinforcing bars, concrete technologists and civil engineers have to interface with electrochemists.

Several developments using new concepts, such as self curing concrete, realkalisation and third generation plasticizers have been developed by multi disciplinary teams.

Scientists have assembled a vast amount of knowledge through a systematic, cyclical process of observation – hypothesis – verifications and predictions and organised them in a systematic way. Whatever be the driving force or motivation for scientific research, either curiosity or human needs, the results of their endeavour are applied to all areas of engineering and technology. The interface between the two communities, scientists and engineers should be open to allow free exchange of information.

(a) This may be by a percolation of scientific principles from the basic scientist to the engineer or applied scientist.
(b) Permeation of ideas from higher Academia to the ground level of business and industry.
(c) Diffusion of concepts across various disciplines to the area of concrete technology.

Supplier – Specifier Interface

The second societal interface is between the supplier and the specifier. While the scientist – engineer team develops new products and systems, the supplier (ready mix concrete industry or construction industry) provides the driving force to take the new technology to the market place. The specifier, most often acting in the interest of the consumer or owner, tries to regulate the supplies or services. This regulatory process may vary from simple resistance (by way of detailed examinations), through procrastination to stubborn refusal.

The performance at this interface will be improved when the driving force provided by the supplier is directed by the specifier in a beneficial direction. Instead of dampening the rate of progress or dumping the innovation, the regulating devices may be directed properly in the following manner (Figure 6).

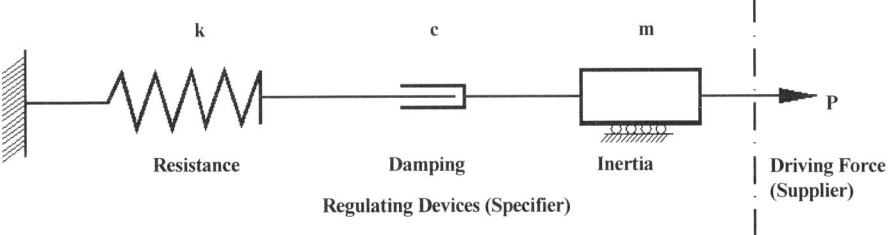

Figure 6 Dynamics of the supplier-specifier interface

(a) Resistance to change: This can be applied when the supplier of a product is emerging as a monopoly.

(b) Delayed response (damping): This may be appropriate when a supplier promotes a product that may be replaced by a better one in the near future.

(c) Refusal to act (inertia): When a supplier makes spurious claims about his product, the specifier may decline to specify the product.

Contractor-Client Interface

The owner of a facility, or a client enters into a contract with a contractor, essentially because it is not worthwhile for the owner to create an establishment and an organisation to carry out construction, which is not in his core business area. In addition to providing the construction services, often at a fixed price, the contractor has to face several risks and the prospect of liquidated damages, penalties and law suits. On the other hand, a client often feels that the should be wary of a contractor who may raise claims on a score of reasons like extended stay, acts of God and denial of access.

Performance at the contractor-client interface can be improved if the owner selects a contractor on the basis of his track record, competence and resources and not on price considerations alone. Alternate dispute resolution mechanisms, such as conciliation and arbitration have displaced bitter and long drawn litigation, but we may occasionally spot a Rip van Winkle, who is unaware of this trend.

Quality surveillance can be an irritant to the superfast contractor, especially working underground or under water where inspection is not possible or convenient after the activity is completed. Quality Systems and Certification to ISO 9000 series are helpful and provide assurance to the owner that the contractor's quality system is adequate and effective. But a word of caution may not be out of place. ISO Certification implies only that a quality system is being operated by the supplier; it does not, however certify that the product or service will necessarily meet the buyer's expectation of quality or performance. For example, a Ready Mix Concrete supplier may not have a water meter in his batching plant and may produce concrete of highly variable strength and yet have ISO 9002 Certification, because all these facts are built into his quality systems, by means of documented procedures and records.

It is a fact of life that the contractor operates in an environment fraught with risks. Longer the duration of the contract, riskier is the enterprise. Fixed price contracts impose the risk of inflation. Unexpected conditions (e.g. foundations) or events (like storms and strikes) pose additional risks which are compensated by insurers to a limited extent only. The performance at contractor-client interface will improve when the owner agrees to share the risk and to award a bonus for good performance instead of imposing a penalty for non-performance!

CONCLUDING REMARKS

Concrete technology and construction industry have changed, in tune with changes in other industry segments. It is necessary to be aware of the change and also the role of technical and societal interfaces in the performance of concrete and of the construction industry.

While the technical aspects of concrete can be understood and developed explicitly, only concerted efforts of scientists, engineers, suppliers, specifiers, contractors and owners will result in better and more durable construction.

REFERENCES

1. AMERICAN CONCERETE INSTUITE. ACI – 318-1977, Building Code Requirements for Reinforced Concrete (Clause 12.2.3). American Concrete Institute, Detroit, Michigan, USA.

2. BEREAU OF INDIAN STANDARDS, SP-23. Hand Book on Concrete Mixes, B.I.S., New Delhi, India, 1982, pp 96-99.

3. MITSUI, K, ZONGJIN L I, LARGE, D A AND SHAH, S P. Influence of Rock and Cement Types on the Fracture Properties of the Interfacial Zone, Chapter 14, Interfaces in Cementious Composition , Ed.., J.C. Maso, E&FN Spon., London, 1993.

4. BENTZ, D P, SCHLANGEN, E, AND GARBOCZI, E J. Computer Simulation of Interfacial Zone Micro Structure and its Effect on the Properties of Cement Based Composites, In Material Science of Concrete, Vol IV, Ed., J. Skalny and S. Mindess. The American Ceramic Society, Westerville, Chio., 1995, p166.

5. ZIMBELMANN, R. A Method for Strengthening the Bond Between Cement Stone and Aggregates, Cement and Concrete Research, Vol.17, No.4, 1987, pp 651-660.

6. SHEN YANG, XU ZHONGZI, XIE PING AND TANG MINGHEE. A New Method of Enhancing Cement – Aggregate interfaces. Part I, Cement and Concrete Research, Vol.22, No.4, 1992. pp 612-620.

7. MINDESS, S. Interfaces in Concrete in Material Science of Concrete, Ed. J. Skalny, The American Ceramic Society, Westerville, Ohio, 1989, p 171.

8. KANDA, T AND LI, V C. Interface Property and Apparent Strength of High Strength Hydrophilic Fibre in Cement Matrix. ASCE Journal of Materials in Civil Engineering, Vol.10, No.1 Feb.1998, pp S-13.

9. GJOREV, O E. Durability ,Chapter 5 in High Performance Concrete Properties and Application, Ed. Shah, S.P. and S.H. Ahmed, McGraw – Hill, New York, 1994, p 155.

10. BENTUR, A. Fiber–Reinforced Cementitious Materials in Material Science of Concrete, ed. J. Skalny, The American Ceramic Society, Westerville, Ohio, 1989, p 273.

11. STROEVEN, P. Some Mechanical Effects of Interface Debonding in Plain Concrete, in Interfaces in Cementious Composites, ed. J.C. Maso, London, E&FN Spon, 1993. pp 187-195.

SURFACE STATE OF DEEP-DRAWING TOOLS IN HYDRAULIC CONCRETE- DESIGNING AND PROPERTIES

A Schwartzentruber

Ecole Normale Supérieure Cachan

J-P Bournazel

LERM

J-N Gacel

SOLLAC

France

ABSTRACT. In the car industry, sheet-steel manufacturers research new solutions to reduce the production cost. Therefore, one proposes to produce these tools in high performance concrete. Nevertheless, in order to increase the number of deep-drawing cycles, it is necessary to improve the surface properties of concrete. Therefore, we have developed a fabrication process to obtain concrete tools with no surface defect, a high hardness and a low roughness. The performances of these tools have been evaluated on an industrial deep-drawing line.

Keywords: Deep-drawing tools, Surface state, Superplasticizers, Surface tension, High performance concrete (HPC).

Arnaud Schwartzentruber is a PhD Student at the Ecole Normale Supérieure, Cachan, France. He is previously of the *agregation* of Civil Engineering and he is Research Engineer at the Laboratory of Research and Studies of Materials (LERM), Arles, France. His main research interests concern, the development and new ways of utilisation of Cementitious High Performance Materials.

Jean-Pierre Bournazel is Research Director of the Laboratory of Research and Studies of Materials (LERM), Arles, France. He is also associate professor at the Ecole Normale Superieure, Cachan, France.

Jean-Noel Gacel is Research Engineer at the Centre of Studies & Development (CED) of the society SOLLAC, Montataire, France. He works on the improvement of forming technique of sheet-steel and on the development of new products based on sheet-steel.

INTRODUCTION

The body of a car contains approximately 300 drawn parts made of steel and most of the parts are formed with several set of deep-drawing tools. Therefore, for a new standard car, the cost of designing and producing deep-drawing tools is an important part of the total investment cast iron tools are profitable for producing more than 1 million parts. On the other hand, the prototype cars, i.e. less than 100 cars, are produced with deep-drawing tools made in polymer resin concrete or metal charged resin. These latter materials are expensive and the number of parts deep-drawn does not exceed some thousands due to the important wear rate of the resin. Consequently, the sheet-steel manufacturers are looking for materials able to deep-draw at a lower cost small or medium series of parts for limited edition cars or trucks for instance.

Indeed, the future orientations of car manufacturers are to renew and modify their models more often in order to adapt them to the demand. Consequently, less and less cars will be produce at a large series, i.e. higher than 1 millions cars.

SPECIFICATIONS OF CONCRETE TOOLS

The first idea was to use hydraulic concrete to produce prototype cars [1,2]. These tools are used less than a hundred times and the dimensional accuracy is not as severe as the tools of large series. At the present time, prototype tools are made in polymer resin concrete (PRC) or, when deep-draw cycles induce important stresses to the tools, in a metal charged resin (MCR). Faced with these materials, hydraulic concrete is easier to handle than PRC, does not need to be machined like MCR and, therefore, the producing delays and costs are reduced. The first deep-drawing tools made in hydraulic concrete has been cast in a metal mould provided by the firm SOLLAC. The tool has small dimensions, i.e. 10 cm each side, in order to limit the size of the formed part. The contact area of this tool contained several shapes (Figure 1) which permit to characterise the ability of a material to the deep-drawing.

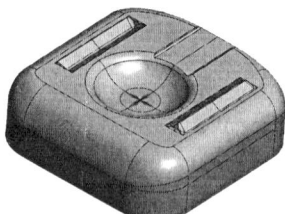

Figure 1 Shape of the contact areas

Numerical simulations realised with the finite elements code Abaqus showed that the maximal compressive stress imposed during a metal forming cycle is 70 MPa [3]. Therefore, a high performance concrete (HPC) with a compressive stress of 90 MPa had been chosen. Nevertheless, this maximal value of the compressive strength can not be generalised because it depends on numerous parameters such as the type of press, the pressure in the die cushions, the cadence, the characteristics of the sheet-steel. The HPC tool has been tested on a industrial deep-drawing line in the laboratory LEDEPP based in the SOLLAC factory at

Florange (France). These tests permit us to deep-draw 1000 parts in which 600 parts were usable [1] and showed that HPC tools can be used to produce prototypes cars. Nevertheless the tests showed that the number of cycles is limited by the wear of the contact areas due to the friction and the abrasion. Moreover, the surface defects of the tools and an heterogeneous wear involve a quick damage of the tools. Actually, the tools have been cast in metal mould covered by a form oil. And, despite all the precautions taken during the casting numerous air bubbles were visible on the tool surface.

Following these tests, the wear and friction properties of the HPC has been determined by a friction test [4]. In the same way, the HPC samples have been cast in a metal mould covered by a form oil. This friction test highlighted different rates of wear which can be linked to the different microstructure present at the surface of the tool. The first zone is 12 µm thick and it is less resistant to wear than the subjacent layer.

These weak properties are due to the local increase of the porosity and to the modification of the microstructure, i.e. nature and orientation of hydrates. This phenomenon is generally called "wall effect", but, at this scale, the wall effect is due to the low packing density of the cement grains against the wall. Another factor which decreases the properties of the surface concrete is the use of form oil. This oil is partially absorbed by the microstructure during the set and weakened the resistance of the concrete near the surface.

HOW TO IMPROVE THE HPC TOOL PROPERTIES?

These first results guided us for improving the performances of HPC tools. The directions of our researches were:

- removing all the defects, i.e. the air bubbles, on the surface of the tools;
- increasing the resistance of the different layers present under the surface;
- ensure a homogeneous wear of the tools;

This article presents the technical solutions we used to resolve the first point.

Removing Air Bubbles

Air is always present in fresh concrete and our aim is not to remove all the air but to prevent the formation or the stabilisation of air bubbles in the surface layer of the concrete. With the intention of determine if air bubbles are more stable against the wall or in the fresh concrete, we study the two systems presented on Figure 2. These systems are composed of 3 phases, a solid phase: the Wall, a liquid phase: the Fresh concrete (Fc) and a gas: the air. The system the more in equilibrium will be the system which need the less energy to be formed.

The stabilisation will depend of the surface tensions between these 3 phases, i e $T_{FC/Wall}$, $T_{Fc/Air}$, $T_{Wall/Air}$. Thus, it is evident that the nature of the wall or the surface product applied on the wall before casting will play an important role on the air bubble distribution against the surface of the mould.

The working hypotheses are:

- same number n of air molecules in the 2 states; air is considered as a perfect gas;
- concrete is regard as a thermostat and all the reactions are isotherm (T_0);
- fresh paste is a homogenous media;

The used notations are as follow:

- P_j, V_j, R_j are respectively the air pressure. the air volume an the radius of the air bubble of the configuration i; P_w is the paste pressure;
- $S_{j\,A/B}$ is the interface surface between the phase A and the phase B of the configuration i;
- α is the contact angle of the air bubble in the configuration 1.

The hypotheses a), b) and c) permit us to write:

$$P_1V_1 = P_2V_2 = nRT_0 = \text{constant} \tag{1}$$

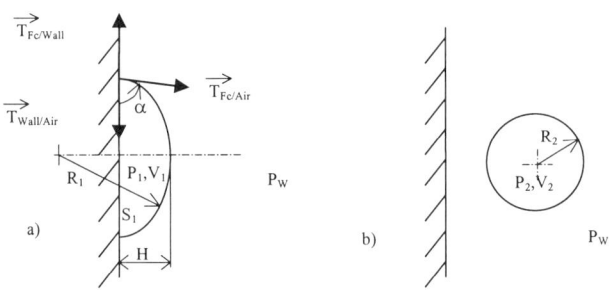

Figure 2 a) an air bubble stuck to the wall, b) an air bubble in the fresh concrete

While the bending radius change between the two configurations, the pressure jump at Air/Fc interface is different. Therefore, the air pressures of the two configurations are different and consequently the air volumes are different (Equation (1)). Nevertheless, the pressure jump is inversely proportional to the radius of the air bubble and it becomes predominant when the bubble diameter is less than 0.5 mm or it is the bubbles of large diameter present at the surface which lead to a quick wear of the concrete tools.

Since we are only interested by the large air bubbles, we can write in first approximation that Vl is equal to V2. This equality leads to a relation between the radius R, and R2 and the angle a (cf. Equation (2)).

$$\lambda(\alpha) = \frac{R_2}{R_1} = \sqrt[3]{\frac{3}{4}\left[(1-\cos\alpha) - \frac{(1-\cos^3\alpha)}{3}\right]} \qquad (2)$$

The Nuclei of an air Bubble at a Liquid/Solid Interface

To form an air bubble against a wall, the fresh concrete needs first to be removed, thus an interface must be delete before creating another. Consequently, the energy (E_1) of the first configuration can be write as follow:

$$E_1 = S_{1FC/Air} T_{FC/Air} + S_{1 Air/Wall} (T_{Air/Wall} - T_{FC/Wall}) + V_1 (P_w - P_1) \qquad (3)$$

In the same way, the energy (E_2) needed to form an air bubble in the volume of fresh concrete is given by the Equation (4).

$$E_2 = S_{2FC/Air} T_{FC/Air} + V_2(P_w - P_2) \qquad (4)$$

Comparison Between the Two Configurations

By using the Laplace's equation and the Young's equation, the difference between E_1 and E_2 is given by the Equation (5). The study of this function shows than an air bubble is more stable against the wall when $\alpha \in [0°;110°]$ and that it is more stable in fresh concrete when $\alpha \in [110°;180°]$. It is the Young's Equation (6) which gives the value of the angle a. The Good-Fowkes model [5] permit to determine the interfacial tension between the wall and the fresh concrete in function of the surface tension of each phase and the interaction between the two phases due to dispersion and polar forces (Equation (7)). Moreover, if we consider that the characteristics of the paste is close to water, we finally obtain the Equation (8).

$$E_2 - E_1 = \frac{2}{3}\pi R_1^2 T_{Fc/Air}\left(2\lambda(\alpha) - 1 + \cos\alpha + \frac{1}{2}\cos\alpha \sin^2\alpha\right) \qquad (5)$$

$$\cos\alpha = \frac{T_{Fc/Wall} - T_{Air/Wall}}{T_{Fc/Air}} \qquad (6)$$

$$T_{Fc/Wall} = T_{Wall/Air} + T_{Fc/Air} - 2\sqrt{T_{Wall/Air}^d \times T_{Fc/Air}^d} - 2\sqrt{T_{Wall/Air}^p \times T_{Fc/Air}^p} \qquad (7)$$

$$\cos\alpha = 1 - \frac{2\sqrt{21.8 \times T_{Wall/Air}^d} + 2\sqrt{53 \times T_{Wall/Air}^p}}{73.8} \qquad (8)$$

The Equation (8) means that one needs to reduce the surface tension of paste and to find a mould material for which the surface tension is high with a polar part higher than a dispersion part.

Nevertheless, a mould with a high surface tension and particularly with a high polar part means that the surface could react chemically with the paste and thus it would be impossible to strip the concrete part from the mould.

Thus, we have to chose a non-polar mould with a low surface tension to prevent any adherence and the only way to prevent the adherence of an air bubble at the surface is to decrease the surface tension of the paste.

Superplasticizers which are usually introduce in concrete mix in order to decrease the quantity of water while maintaining a good workability are polyeletrolytes with hydrophilic parts and hydrophobic parts. Consequently, a molecule of superplasticizer can be considered as a long chain of surface-active molecules.

Nevertheless, the size of these macromolecules limits their movement facilities and they are less efficient than surface-active molecules to reduce the surface tension of water. In order to determine their surface-active properties, we choose four different superlasticizers and we measure the fall of the surface tension of water in function of their concentration.

We use the Wilhelmy slide method [6] with a piece of platinum foil. The reference liquid was distilled water and after each measure, the platinum foil has been clean several times to remove the superplasticizer. Indeed, a few molecules present at the surface the foil can modify notably the measures. Moreover, all measures have been repeated three times. The results are presented on the Figure 3 and show that the effects of superplasticizers on the surface tension depend strongly on its nature. Polynaphtalene sulfonate (PNS) or polymelamine sulfonate (PMS) which are the first generation of superplasticizers have no surface-active properties. Conversely, the new generation of superplasticizers like polyacrylate (PA) or modified polyphosphate (MPP) have clearly surface-activesproperties.

Similar results have been found by Kinoshita *et al.* [7]. They showed that PNS do not decrease the surface tension of water while high-range water-reducing agent based on water-soluble graft copolymer decrease strongly the surface tension of water. These results can be related to the different dispersion mechanisms. Actually, for the first generation of superplasticizers, it is principally electrostatic repulsive forces, trough their sulfonate ends, which guarantee the dispersion of the different particles while for the new superplasticizers, it is principally steric repulsive forces trough their long hydrophilic chains graft which are at the origin of the good workability of the fresh concrete.

In order to verify the spreading facilities, we measure the contact angle of a water drop with different concentrations of superplasticizers on a surface covered by a non-polar painting. The polar and disperse parts of the painting surface tension is given by the Equation (9) and the results are showed on the Figure 4.

$$T_{P/Air} = T_{P/Air}^{d} + T_{P/Air}^{p} = 38.7 + 5.4 = 44.1 \text{ mJ/m}^2 \tag{9}$$

These measures show that even a low concentration of MPP in water makes the surface which was initially non-polarized, i.e. hydrophobic, hydrophilic and consequently improve the sDreadin~ of water on the surface. While for the old generation of superplasticizers, i.e. PNS and PMS, there is no decrease of the contact angle. The PA has intermediary spreading properties since its contact angle is slightly lowered. These results are in good agreement with the measures of the surface tension.

Concrete Deep Drawing Tools 117

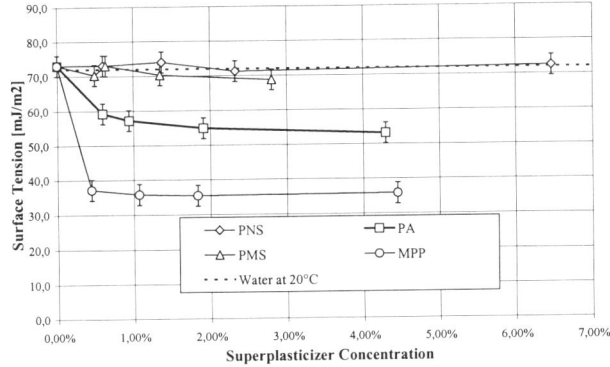

Figure 3 Evolution of the surface tension of water in function of the concentration of four different superplasticizers

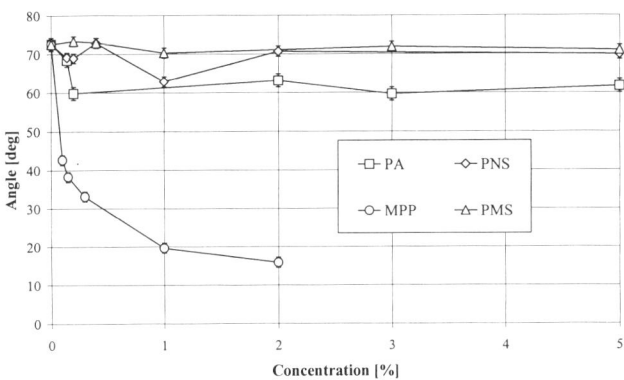

Figure 4 Evolution of the contact angle of a water drop with different concentrations of superplasticizers

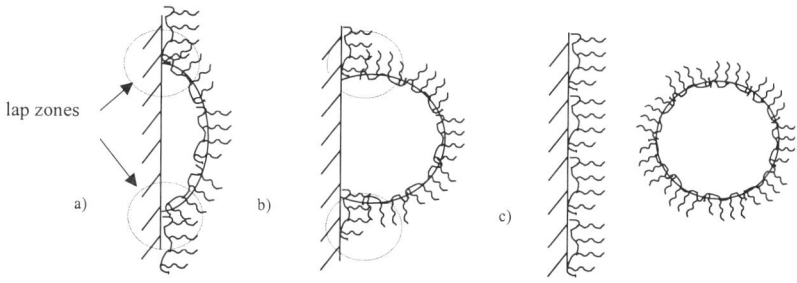

Figure 5 Mechansim of detachment of air bubble

Indeed, the spreading mechamsm is linked to the surface tension since the hydrophobie parts of the superplasticizer stand against the non-polar surface whereas the hydrophilic part will turn towards the fresh concrete. By this way, a layer of superplasticizer is created against the wall. The surface of the mould is covered by a layer of hydrophilic parts and can be considered hydrophilic. The superplasticizer layer improve the spreading of the water drop and replaces the form oil. The surface-active properties of some superplasticizers have other advantages. The superplasticizer molecules are spread out around the air bubbles with the hydrophobic part on the air side and, if an air bubble is still stick to the wall, the repulsion forces between the molecules present at the lap zone will promote the detachment of bubbles (Figure 5). Furthermore, the layer of superplasticizer present against the surface of the mould will give a plastic aspect to the concrete surface, i.e. glassy and shiny.

Finally, in order to obtain a concrete surface without any defect, you have to chose a mould made in a non-polar material and to introduce in the mix a superplasticizer able to reduce strongly the surface tension of water. By this way, you do not need to apply form oil on the mould. Nevertheless, in order to facilitate the detachment of the more adherent air bubbles, the use of a vibrating table is recommended. Indeed, the mould surface is never perfect and a scratch or shape with a small bending radius can increase the adherence of air bubbles However, with this method, there is an important capillary adherence between the concrete part and the mould. Indeed, during hydration, numerous small capillaries are formed against the surface of the mould due to the consumption of water by the hydrates. These capillaries are not fill with air and thus a depression is created. Consequently, the adherence force is proportional to the moulded surface area and for large concrete part, if you want to save the mould, it is necessary to provide an injection device of air between the mould and the concrete parts

TEST OF THE DEEP-DRAWING TOOLS UNDER PRESS

HPC tools produced with the above process has been evaluated on a, quasi-industrial deep-drawing line. The surface of the tool had initially no defect, it was smooth and shiny. This tool was able to deep-drawn 2000 usable parts. The wear of the surface was homogenous and moreover the material was able to correct the defects which can appear on the surface of the tool during the deep-drawing cycles, i.e. appearance of air bubbles entrapped in the second or third layer. Instead of the defects grown in a prejudicial manner, they were polished during the following deep-drawing cycles and after about a hundred cycles, the defects have disappeared.

CONCLUSIONS AND PERSPECTIVES

The fabrication process developed in this article perm~t us to increase the number of deep-drawing parts produced with a HPC from 600 to 2000 without increasing strongly the compressive strength of the concrete due to the increase of the silica fume content. The solution lies in the use of a mould with a low surface tension without any form oil, the introduction of a superplasticizer with surface-active properties, the use of an optimal quantity of silica fumes in order to reach the maximal packing density, a good vibration of the mould and preventing the desiccation of the surface layer during at least three days.

However, we see the limits of an HPC tools because its performances depend on the hardness of the paste which can not be more improve without increasing the quantity of cement or applying a thermal cure. Nevertheless, the Very HPC (VHPC) are more brittle than HPC and the successive shocks can lead to a brutal rupture of the tools. At the present time, fibers are introduced in the mix in order to improve the ductility of these materials. Nevertheless, metallic fibers can scratch the sheet-steel or fibers can be pull out. Moreover, the new VHPC contains usually siliceous grains or powders which makes the concrete hardly machinable after being molded which imply some modifications in the conception process.

REFERENCES

1. SCHWARTZENTRUBER, A, BOURNAZEL, J-P, GACEL, J N, Hydraulic Concrete as a Deep-Drawing Tool of Sheet-Steel, CCR, to be published.

2. ROGER, V, BOURNAZEL, J-P, Internal Report, Contract SOLLAC/LMT, July 1995.

3. BURLAT, M, Analyse' mecanique et tribologique de l'emboutissage: application aux outils fabriques par combinaison de differents materiaux, Thesis of the INSA de Lyon, France, presented the 20 April 1998.

4. SCHWARTZENTRUBER, A, BOURNAZEL, J-P, GACEL, J-N, Conception d' outils d'emboutissage en beton hydraulique , Proceedings of the 1st International Meeting "Material Science and Concrete Properties", Toulouse, France, pp. 349-356, March 5-6, 1998.

5. FOWKES, F M. Adv. Chem., 43, 99 (1964).

6. WILHELMY, L, Ann. Phys., 119, 177 (1863).

7. KINOSHITA, M, YONEZAWA, T, YUKI, T. Chemical structure and performance of a new type High-Range Water-Reducing Agent for Ultra High Strength Concrete, JCA Proceedings of Cement & Concrete, pp. 196-201, 47, (1993).

EFFECT OF INCLUSION OF DENSE AGGREGATES ON THE PROGRESS OF THE HYDRATION PROCESS AND STRENGTH DEVELOPMENT IN MORTAR AND CONCRETE

K van Breugel

Delft University of Technology

E A B Koenders

Heerema International

Netherlands

ABSTRACT. Mortars and concretes made with dense aggregate are generally considered as three phase materials. The three phases are the cement paste, the aggregate and the matrix-aggregate interfacial zone. In concrete made with dense aggregate the interfacial zone is generally more porous than the bulk paste and is considered the weakest link. As a consequence of this the strength of mortar and concrete is often lower than the strength of the paste of which the concrete is made. This in spite of the fact that the degree of hydration of the cement in concrete has been reported to be higher than that of plain paste made with the same water/cement ratio.

Why and to which extent the presence of dense, non-absorbing aggregate particles may affect the progress of the hydration process was investigated as part of an extensive study on "ribbon paste hydration". The ribbon paste represents the layer of paste between aggregate particles. Ribbon paste hydration was analysed with the simulation program HYMOSTRUC, developed at the TU-Delft. It is shown how with progress of the hydration process water from the water-rich interfacial zone is transported to the bulk paste, promoting hydration of the latter. The resulting degree of hydration of the ribbon paste is compared with the degree of hydration reached in pure (plain) paste, whereby the former was found to be higher than the latter. Results are compared with data reported in literature. The study gives insight in factors which determine the difference in strength between paste, mortar and concrete and shows how numerical simulation programs can be used as a strategic research tool for detailed analysis of specific hydration features.

Keywords: Hydration, Microstructure, Paste, Aggregate, Interfacial zone, Moisture transport, Numerical simulation, Strength, Concrete.

Dr Ir K van Breugel PdD civil engineering, is senior researcher and lecturer at Delft University of Technology. His research topics are early age concrete, design for imposed deformations and concrete structures for environmental protection.

Dr Ir E A B Koenders civil engineering, got his PhD from Delft University of Technology in 1997. During his PhD study he was involved in a BRITE EURAM project on high strength concrete. His thesis is devoted to the numerical simulation of volume changes in hardening concrete. At present he is employed at Heerema International, Rotterdam, The Netherlands.

INTRODUCTION

The properties of hardening and hardened concrete strongly depend on the microstructure of the material. This microstructure is made up of a solid component, consisting of unhydrated cement and hydration products, or gel, and a pore system that is partly filled with water and partly with water vapour. The properties of the microstructure depend on the initial water/cement ratio, the degree of hydration and the curing temperature [1,2]. In the case of concrete, the microstructure is not homogenous throughout the paste, but exhibits "porosity gradients". The reason for porosity gradients is the non-homogenous distribution of water in the concrete. Depending on the fineness of the cement, the water/cement ratio and the properties of the aggregate, a matrix-aggregate interfacial zone is formed of which the water/cement ratio is different from the nominal value. The course of the water/cement ratio with increasing distance from the aggregate surface is shown schematically in Figure 1. As is known from hydration tests on pure cement pastes, the water/cement ratio has an important effect on the rate of hydration and on the microstructure that is formed. Knowing this, it is obvious that the microstructure of the interfacial zone is different from that of the bulk paste.

The formation of a porous interfacial zone has been mentioned as a reason why hydration of cement in a concrete mixture proceeds faster than in a pure paste. Whether this can really be considered a satisfactory explanation is not immediately clear. In this respect it must be realised that the accumulation of water near the surface of dense aggregate particles is associated with a decrease of the water/cement ratio in the bulk paste. The rate of hydration in the bulk paste will, therefore, be lower than in the case of a pure paste with the nominal water/cement ratio. Whether the presence of water in the porous interfacial zone is a sufficient condition for an increased rate of hydration of "concrete paste" compared to "pure paste" is, therefore, still debatable. Another important question concerns the mechanism through which water from the water-rich interfacial zone could be transported to the bulk paste. This bulk paste can only benefit from the presence of nearby-water if there is a driving force that causes transport of water from water-rich zones to the relatively dry bulk paste. Tracing these mechanisms and quantification of them is in the focus of attention of this paper. In order to be able to quantify moisture transport within the hardening paste its microstructure and permeability must be known. Quantification of the microstructural development and permeability in hardening paste shall, therefore, be addressed in more detail in this paper.

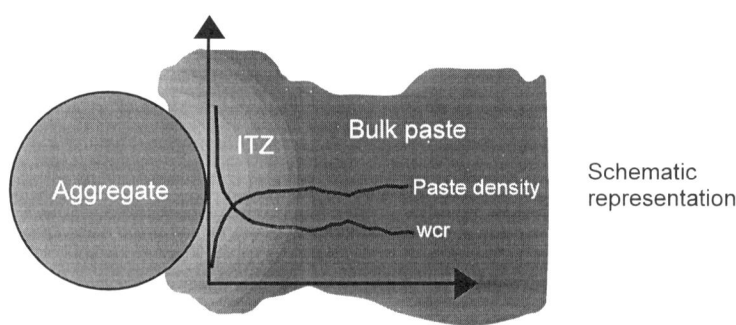

Figure 1 Schematic representation of matrix-aggregate interfacial zone [3]

HYDRATION OF PURE PASTE

Models for Hydration and Microstructural Development

One possible way to investigate possible transport of moisture within a hardening "concrete paste" is by numerical simulation. These numerical simulations concern quantification of the hydration process and the associated microstructure. For numerical quantification of hydration and microstructural development only a few models are available [5,6]. In this study the simulation program HYMOSTRUC [2] has been used. In this model hydrating cement grains are considered as gradually expanding spheres. Expansion occurs as the result of precipitation of reaction products on the outer surface of the hydrating grains. The expanding spheres will make contact with adjacent particles as shown schematically in Figure 2. In the model the process of external growth and formation of contacts with adjacent particles has been developed in a mathematical series [2]. The model keeps track of microstructural changes and changes in the amount of capillary water and the relative humidity in the pore system.

The original HYMOSTRUC model, in which cement grains of the same size are positioned at equal distances, has been modified by Koenders [3] in order to allow for the actual randomness of the cement grains in the paste. An arbitrary space is filled randomly with cement particles, starting with the biggest particles and subsequently with the smaller ones. The position of each particle is defined by its co-ordinates as indicated, by way of example, in Figure 3.

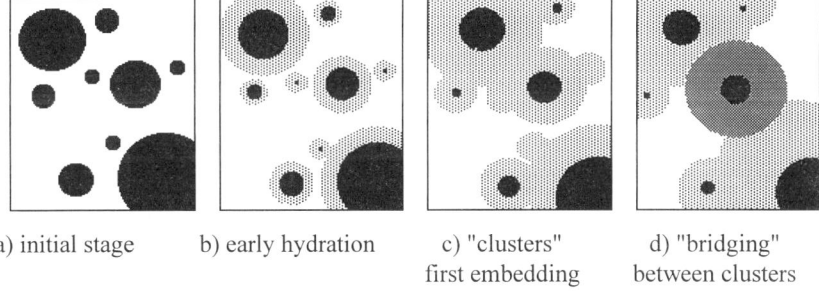

a) initial stage   b) early hydration   c) "clusters"   d) "bridging"
                                          first embedding   between clusters

Figure 2 Formation of a microstructure in hardening paste as considered
in the simulation program HYMOSTRUC [3]

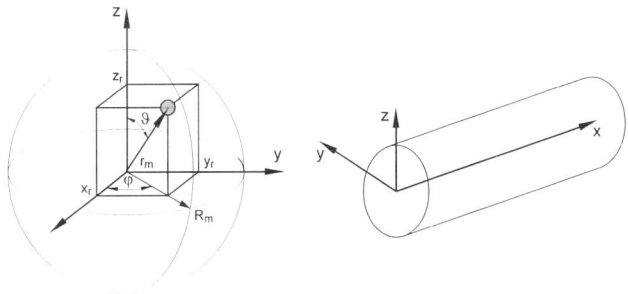

Figure 3 Random particle distribution in spherical and tubular paste volume or [3]

For a spherical paste volume, with a diameter of the sphere of 200 μm, the result of a numerical simulation of the generated microstructure is shown in Figure 4. The simulation concerns a paste made with a cement with Blaine fineness 420 m^2/kg and a water/cement ratio 0.3. Two stages of the hydration process are presented, viz. a degree of hydration α = 0.1 and α = 0.5.

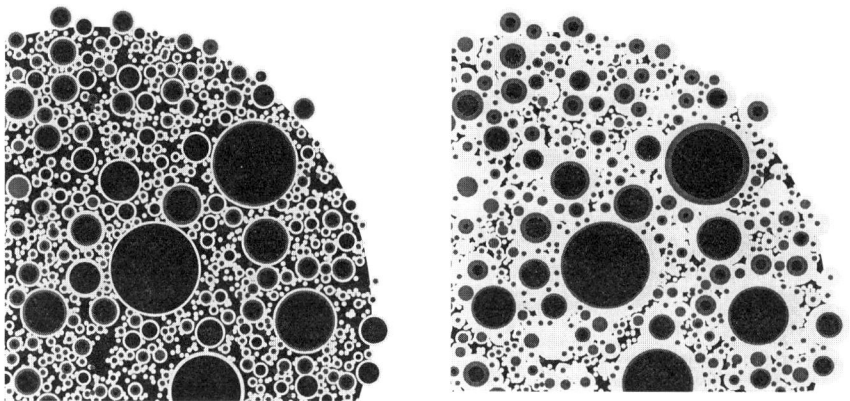

Figure 4 Simulated microstructures for paste. Blaine 420 m^2/kg, wcr = 0.3. Degree of hydration α = 0.1 (left) and α = 0.5 (right) [3]

Pore Structure and Permeability

From the numerically generated microstructure information about the pore size distribution can be deduced by "pixel counting". Such a counting procedure is shown in Figure 5, where the evolution of the pore size distribution with time is presented for an arbitrary chosen paste. Since the HYMOSTRUC program keeps track of the amount of capillary water in the pore system and knowing that this capillary water is accumulated in the smaller pores, it is possible the determine the size of the pores that are still completely filled with water. The remaining part of the pore system, consisting of bigger pores, will be filled with water vapour.

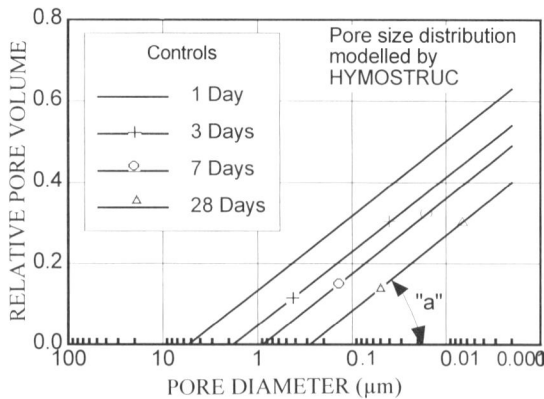

Figure 5 Simulated pore size distribution in hardening cement paste

Based on the work Setzer [7] and Gertis et al. [8], a procedure has been developed to determine the water and gas permeability for a given pore size distribution with a certain degree of saturation. Results of water permeability calculations are presented in Figure 6. In this figure, water permeability is given as a function of the degree of hydration for pastes made with Portland cement, Blaine value 420 m²/kg, and water/cement ratios 0.3, 0.4 and 0.5, respectively. Bearing in mind signi-ficant scatter in experimental data, the calculated values are in reasonable agreement with the values obtained experimentally

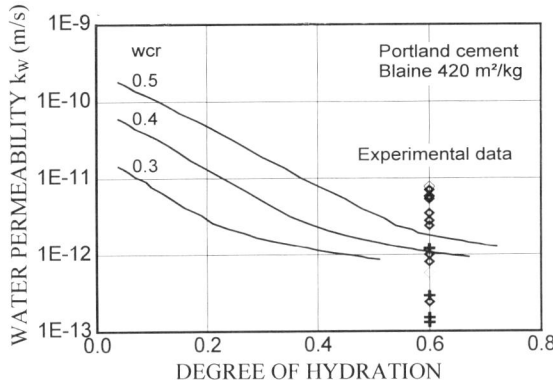

Figure 6 Calculated water permeability as a function of the degree of hydration [3]

Self Desiccation of Plain Paste – Relative Humidity

In case of sealed curing of concrete the relative humidity in the pore system gradually decreases with progress of the hydration process. This is known as self-desiccation. In paste with a low water/cement ratio, e.g. wcr = 0.3, the relative humidity may drop down to 70%. In Figure 7 calculated values of the relative humidity are presented as a function of time for three different water/cement ratios. These calculated values are in reasonable agreement with measurements of other authors [8-10].

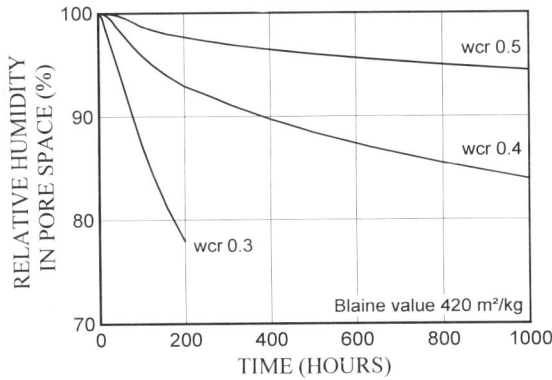

Figure 7 Relative humidity in three hydrating cement pastes as a function of time [3]

FORMATION OF INTERFACIAL ZONE

Stereological Wall Effect as a Cause of the Formation of an Interfacial Zone

In real concrete we are not dealing with pure paste, but with "ribbon paste". Ribbon paste is defined as the paste between aggregate particles. As discussed previously the properties of cement paste in real concrete differ from those in pure paste because of the non-homogenous distribution of the water in the paste. In [3] the hydration properties of ribbon paste are compared with those of pure paste. These differences originate from the differences in the initial water content in the matrix-aggregate interfacial zone and the bulk paste. The variable water content over the thickness of the ribbon was considered to be caused by a stereological "wall effect". Due to this wall effect the water content of the paste just adjacent to the surface of the aggregate particles is higher than at greater distance from the aggregate surface. Schematically this was shown already in Figure 1. This stereological wall effect was simulated numerically by filling a tubular space, representing the ribbon paste, randomly with cement particles. Close to the surface of the aggre-gate the local water/cement ratio of the paste is very high but decreases rapidly with increasing distance from the aggregate surface. The reverse is the case with the amount of the cement in the paste (i.e. the "paste density"). In Figure 8 the calculated local water/cement ratio and paste density are presented as a function of the distance from the aggregate surface for a paste made with cement with a Blaine fineness 420 m^2/kg. The average nominal water/cement ratio was 0.30. Due to the less dense packing of cement particles in the interfacial zone the local water/ cement ratio at the aggregate surface substantially exceeds the nominal mean value.

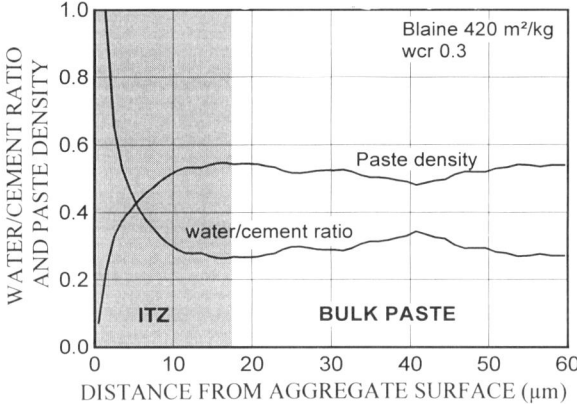

Figure 8 Local water/cement ratio versus distance from aggregate surface for pastes made with cement of different fineness (Blaine fineness), wcr = 0.30 [3]

Numerical Simulation of Hydration-Induced Moisture Transport in Ribbon Paste

With the numerical simulation program HYMOSTRUC the progress of the hydration process over the ribbon thickness could be investigated. For that purpose the ribbon is subdivided in small elements (Figure 9). Each of these elements has a specific initial water/cement ratio. For each of these elements the simulation program HYMOSTRUC calculates the evolution of the hydration process. Due to the great differences in initial water/cement ratio of the elements, the rate of hydration will be different as well. Even more important is the fact that the porosity and the state of water in the subsequent elements will develop differently. In the elements representing the bulk paste a denser pore structure is formed in which less water is left than in the interface paste. The relatively low water content in the bulk paste will trigger a moisture flow from the water-rich interfacial zone towards the bulk paste. Mathematical details of the transport mechanism are described in [3]. This hydration-induced moisture flow affects the rate of hydration of the bulk paste compared to the situation that moisture transport would be ignored. The moisture flow will depend on the hydration-induced moisture gradients and the permeability of the paste.

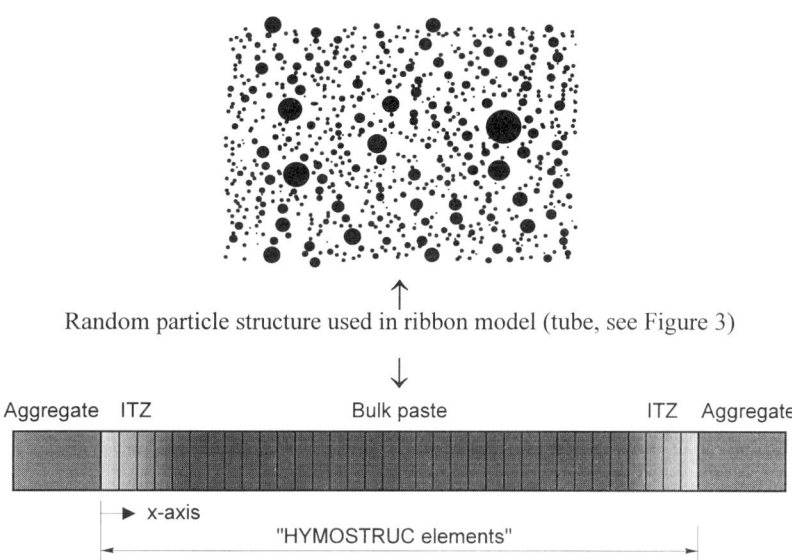

Figure 9 Proposed model for calculation of hydration and microstructure in the interfacial zone

RESULTS OF NUMERICAL SIMULATIONS

Permeability of Ribbon Paste

The water-rich interfacial zone will result in a higher porosity and a higher permeability of this zone compared to the porosity and permeability of the bulk paste. This even though most of the cement present in this zone will be converted into hydration product very soon. The proposed simulation model has the potential to quantify the porosity gradients and the variation of the water and gas permeability within the ribbon paste. For a paste made with cement with Blaine = 420 m^2/kg and wcr = 0.3, the calculated permeability for water and gas over the ribbon thickness of 100 μm after 10 and 100 hours hydration is presented in Figure 10. The high permeability at the aggregate surface decreases steeply to low values in the bulk paste.

Relative Humidity in Ribbon Paste

From Figure 10 it is obvious that the bulk paste is denser than the interface paste and that the effective water/cement ratio of the bulk paste is lower than that of the interface paste.

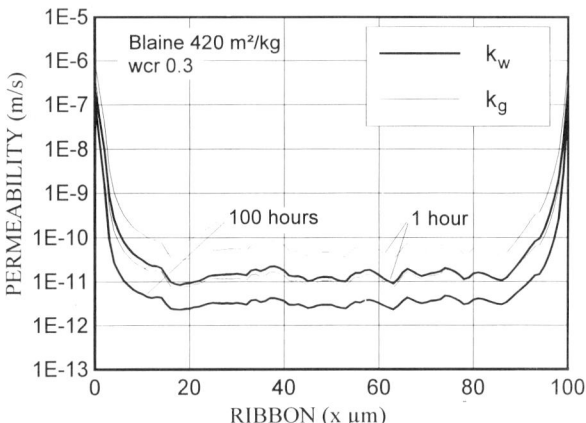

Figure 10 Water and gas permeability of ribbon paste for a paste with nominal wcr = 0.3 [3]

Consequently, the relative humidity and the vapour pressure will vary within the paste. If no moisture transport within the ribbon paste is considered, the relative humidity after 10 and 100 hours hydration will get values as shown in Figure 11 with the thick solid lines. If moisture transport from the water-rich interfacial zone to the relatively dry bulk paste is taken into account, the drop of the relative humidity in the bulk paste will be less, as indicated with the thin solid lines in Figure 11.

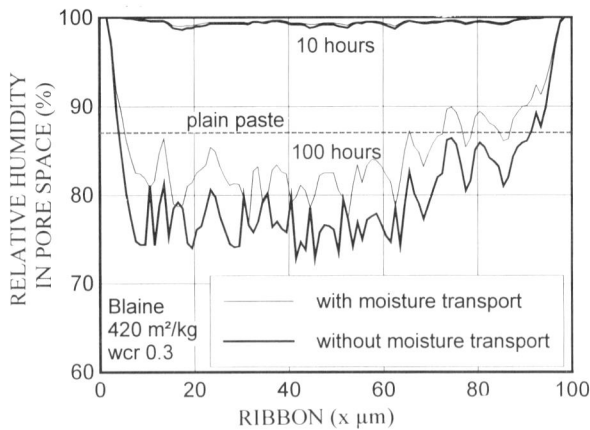

Figure 11 Calculated effect of moisture transport on relative humidity in ribbon paste and pure paste with wcr = 0.3 [3]

In the figure the relative humidity of the pure paste is indicate as well. After 100 hours the relative humidity in a pure (plain) paste would be 87%. In case of ribbon paste the relative humidity would vary between about 100% at the aggregate surface and 73-78% in the bulk paste. This higher relative humidity in the bulk paste will have promoted the hydration process. The calculations reveal, however, that even though the degree of hydration in the bulk paste has increased due to the supply of water from the water rich zone, the drop of the relative humidity in the bulk paste is less than in the case that no moisture transport is considered

Degree of Hydration in Ribbon Paste

In Figure 12 the calculated degree of hydration in ribbon paste is presented after 10 and 100 hours hydration for a paste, water/cement ratio of 0.30 and Blaine fineness of 420 m^2/kg. The thickness of the ribbon is 100 μm. The degree of hydration after 100 hours hydration is presented for a calculation with and without taking into account the effect of moisture transport from the interfacial zone to the bulk paste. For this particular mixture a small increase of the degree of hydration in the bulk paste is found. Compared with the degree of hydration in a plain paste with the same nominal water/cement ratio, the ribbon paste has reached a higher degree of hydration. This numerical result is in good agreement with the findings of Roth [12], who mentioned a 10% higher degree of hydration in "concrete paste" than in pure (plain) paste.

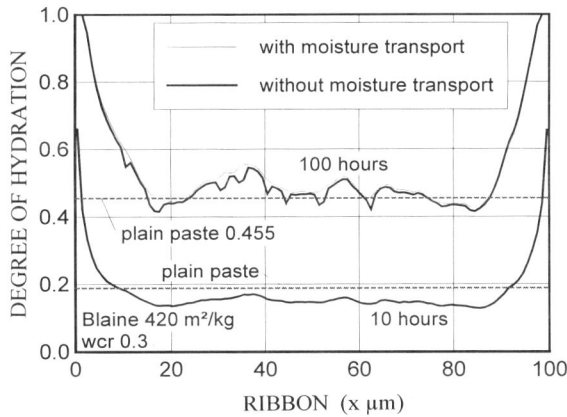

Figure 12 Calculated effect of moisture transport on degree of hydration over the thickness of a ribbon. Blaine = 420 m^2/kg, wcr = 0.30 [3]

DISCUSSION AND CONCLUDING REMARKS

The strength of cement paste is generally higher than that of mortar and concrete. The compressive strength of paste has been reported about 30% higher than mortar strength and up to 50% higher than the strength of ordinary concrete [2]. Obviously the inclusion of rigid aggregate particles in the matrix weakens the material. Based on theoretical considerations Fagerlund [13] states that the strength of concrete and mortar can never be as high as that of pastes. Walz, however, found that under certain conditions, for example in the case of a mix with a low w/c ratio, the strength of concrete could exceed the strength of the paste [14]. Also the use of silica fume can result in strengths higher than the paste strength [15].

An explanation for concrete and mortar strengths lower that the strength of the paste generally goes back to the strength reducing effect of the weaker interfacial zone. By using a low wa-ter/cement ratio or by adding silica fume to the concrete, the quality of the interfacial zone can be improved. The strength of mortar and concrete will hence become closer to the strength of the paste. It is not easy to understand, however, how an improvement of the interfacial zone can ever result in a concrete strength higher than the strength of a paste, unless we consider a possible increase of the degree of hydration due to inclusion of dense aggregates. In this paper the potential of increased hydration of "concrete paste", i.e. ribbon paste, compared with "pure paste" has been illustrated by numerical simulation. Increased hydration due to the presence of a water-rich interfacial zone, as assumed by different authors, was also found by numerical simulation. The simulations offer a better insight in possible mechanisms which contribute to the development of the materials properties. At the same time it is emphasised that this type of simulations is still in a stage of development. Even in this stage, however, they help to understand the mechanisms behind macroscopic phenomena. The effect of inclusion of rigid particles, as dealt with in this paper, is only one of the effects which has our interest. The next step is to consider the inclusion of saturated lightweight aggregates on the rate of hydration and on the effect of these "water reservoirs" on the development of the relative humidity and associated deformational behaviour of hardening pastes. Research of this kind is now being performed in the framework of an international research project EuroLightcon, a european project on the application of lightweight aggregate concrete.

REFERENCES

1. KJELLSEN, K O, DETWILER, R J, GJORV, O E. Development of microstructures in plain cement pastes hydrated at different temperatures. Cement & Concrete Research, Vol. 21, 1991, pp. 179-189

2. BREUGEL, K, van. Simulation of hydration and formation of structure in hardening cement-based materials. PhD, Delft, 1991, p. 295.

3. KOENDERS, E A B. Simulation of volume changes in hardening cement-based systems. PhD, Delft, 1997, p. 171.

4. LOCHER, F W. Die Festigheit des Zements. Beton Heft 7, pp. 247-249, Heft 8, 1976, pp. 283-285.

5. GARBOCZI, E J, BENTZ, D P. Computer-based models of the microstructure and properties of cement-based materials. 9th Int. Conf. Chem. Cem., Vol. VI, 1992, pp. 3-15.

6. JENNINGS, H, et al. The modelling of Microstructure and its potential for studying transport properties and durability. NATO ASI Series E, Vol. 304, 1996, 558 p.

7. SETZER, M J. Einfluss des Wassergehalts auf die Eigenschaften des erhärteten Betons. Deutscher Ausschuss Für Stahlbeton DAfSt H. 280, 1978, pp. 43-79.

8. GERTIS, K, KIESSL, K, WERNER, H, WOLFSEHER, U. Hygrische Transportphanomene in Baustoffen. Deuscher Ausschuss fur Stahlbeton, Heft 258, 1976, pp. 110.

9. WITTMANN, F H. Grundlagen eines Modells zur Beschreibung charakteristischer Eigenschaften des Betons, Deutscher Ausschuss fur Stahlbeton, Heft 290. 1977.

10. BAROUGHEL-BOUNY, et al. Proceedings of the 4th Int. Symposium on Utilization of High Strength /High Performance Concrete. Paris, 1996, pp. 451-461.

11. JENSEN, O. Thermodynamic limitation of self desiccation. Cement & Concrete Research, Vol. 25, 1995, pp. 157-164.

12. ROTH, W. Heat of hydration and degree of hydration of Portland cement. PhD, Aachen, 1970

13. FAGERLUND, G. Relations between the strength and the degree of hydration or porosity of cement paste, cement mortar and concrete. Seminar on hydration of cement, Copenhagen, 1987, 56 p.

14. WALZ, K. Beziehungen zwischen Wasserzementwert, Normfestigkeit des Zements (DIN 1164) und Betondruckfestigkeit. Beton, Nr. 11, 1970. pp. 499-503

15. SCRIVENER, K L, BENTUR, A, PRATT, P L. Quantitative characterization of the transition zone in high strength concrete. Advances in Concrete Research, Vol. 1, No. 4, 1988, pp. 230-237.

EVALUATION OF BOND BETWEEN REINFORCING STEEL AND HIGH STRENGTH CONCRETE ACCORDING TO INTERNATIONAL STANDARDS

I R de Almeida

R H F de Souza

M D Felicio

Universidade Federal Fluminense

Brazil

ABSTRACT. Bond between reinforcing steel and high strength concrete was evaluated, in a two-part study. In the first part, bond tests were executed for a group of ten types of concrete. The conclusions were that in general cement replacement reduced the bond of concrete to steel (silica fume was an exception); utilization of superplasticizers increased concrete-steel bond; bond varied inversely with W/C ratio and directly with compressive strength of concrete; and bond is strongly influenced by the type of cementing material. In this work, as regards the second part of the study, a comparative analysis was carried out between required anchorage lengths calculated based on the experimental data and on the formulae of two international standards. The main conclusions are that CEB/FIP MC 90 and NS 3473 E overestimate the theoretical values, although the ones calculated by the Norwegian Standard are more similar to the experimental data.

Keywords: High strength concrete, Bond, Anchorage Length, Superplasticizers, Condensed silica fume, Fly ash, Natural pozzolan, Admixtures, Reinforcing steel bars.

Prof Dr Ivan Ramalho de Almeida PhD in Civil Engineering (Instituto Superior Técnico da Universidade Técnica de Lisboa, Portugal, 1990), is Full Professor of Construction Materials at the Universidade Federal Fluminense, Niterói, and Fellow Professor of High Strength Concrete at COPPE/UFRJ, Rio de Janeiro, Brazil. He is also Associate Member of ACI Committees 201 and 363, and is also a researcher on high strength concrete with chemical and mineral admixtures.

Profa Dra Regina Helena Ferreira de Souza PhD in Civil Engineering (Instituto Superior Técnico da Universidade Técnica de Lisboa, Portugal, 1990), is Full Professor of Structural Stability at the Universidade Federal Fluminense, Niterói, Rio de Janeiro, Brazil. She is a researcher on structural repair and restoration.

Enga Márcia Dacache Felício MSc in Civil Engineering (Universidade Federal Fluminense, Niterói, Rio de Janeiro, Brazil, 1996) is research assistant at the same university, and her M.Sc. thesis was on Bond Tests for Reinforcement Steel.

INTRODUCTION

Bond between concrete and steel bars in reinforced concrete structures is governed by several factors. In pull-out tests, in general, concrete-steel bond is influenced by compressive and tensile strengths of concrete, concrete shrinkage and bleeding, as well as by the type of superficial deformation, diameter, and chemical composition of steel.

Condensed silica fume (CSF) seems effective in increasing concrete-steel bond [1, 2], and the same occurs with superplasticizers [3, 4].

The first part of this work, already concluded and published [5], was intended to assess these aspects in high strength concrete. In the second part, the anchorage lengths deduced by experimental data and theoretical code formulae were compared. The code equations evaluated are the CEB/FIP MC 90 [6] and the Norwegian Standard [7], two documents applicable to the analysis of high strength concrete structures.

EXPERIMENTS AND RESULTS

In the first part of this work [5], besides a comprehensive bibliographical review on the subject, experiments were carried out on bond between reinforcing steel and 10 different concrete mixes with compressive strength ranging from 60 to 110 MPa at 28 days age. Table 1 shows the composition and main characteristics of these concrete mixtures.

The mixtures studied were divided in two groups of 5 concrete each. The first consisted of a reference mixture with 500 kg/m^3 of cement and four mixtures with 10% of cement replaced by mineral admixtures, such as CSF from Norwegian and Portuguese sources, a Portuguese pulverized fuel ash (PFA) and a natural pozzolan from Cape Verde. The second group consisted of concrete with the same cementing materials as the one of the first group. A superplasticizer was nevertheless added to reduce the mixing water, at a constant concrete workability (40 ± 10 mm slump). Concrete compositions were thus obtained with water-to-cementing materials ratios (W/C) ranging from 0.24 to 0.42. Superplasticizer reduced water content by 31 - 34%.

Tests were carried out following the ASTM Standard C 234 [8] but the geometry of the test specimens was based on recommendations given in the documents of RILEM 7.11.128 [9] and RILEM/CEB/FIP [10].

Test specimens were concrete cubes with 200 mm sides (10 times the diameter of the bar), each with a steel bar incorporated along its axis in a defined length (five times the diameter of the bar). The effective encasement height of the bar corresponds only to the half-height of the specimen. In the other half the bar does not adhere, in order to reduce the influence of the disturbed area that forms close to the bearing plate. The non-adhering part in the concrete was provided by a rigid smooth plastic bushing. The bar was loaded at its longer end by a tensile force whereas at the other end, a device for measuring the displacement between steel and concrete was located, but remained free from stress. The relation between the tensile force and the relative displacement between steel and concrete was measured. The load was increased up to failure of the adhesion. Figure 1 shows a schematic representation of the test setup.

Table 1 Compositions and some characteristics of concrete

COMPONENTS (kg/m³)	CONCRETE MIX									
	B1	B2	B3	B4	B5	B6	B7	B8	B9	B10
Coarse Aggregate	1088	1147	1154	1139	1018	1016	1139	1131	1071	1032
Fine Aggregate	645	681	685	676	602	600	676	670	635	611
Cement	500	500	450	450	450	450	450	450	450	450
Water	182	122	125	120	210	212	122	126	182	201
Superplasticizer (l/m³)	-	15	15	15	-	-	15	15	-	-
CSF A	-	-	50	-	50	-	-	-	-	-
CSF B	-	-	-	50	-	50	-	-	-	-
PFA	-	-	-	-	-	-	50	-	50	-
Pozzolan	-	-	-	-	-	-	-	50	-	50
CHARACTERISTICS										
Water Reduction, in relation to B1, (%)	0	-34	-31	-34	+15	+17	-33	-31	0	+10
W/C ratio	0.36	0.24	0.25	0.24	0.42	0.42	0.24	0.25	0.36	0.40
Compressive Strength in 15 cm cubes (MPa)	67.2	83.7	106.2	105.4	74.9	72.4	86.7	90.7	66.0	61.2
Flexural Tensile Strength (MPa)	6.0	8.1	10.2	10.9	7.8	7.9	7.7	8.6	5.8	5.6
Maximum Bond Stress (MPa)	21.5	27.5	31.2	31.6	24.0	24.8	27.6	26.0	24.0	23.0

- Coarse aggregate is a crushed granite, with 25 mm maximum size, and fine aggregate is a natural siliceous sand, 3.40 fineness modulus.
- Cement is Portland, ASTM type I, and superplasticizer is naphthalene based.
- Condensed silica fume A is a commercial product of Norway, a powder in water suspension, and condensed silica fume B is a Portuguese by-product, in powder form.
- Pulverized fuel ash is a Portuguese by-product (ASTM type F), in powder form, and Pozzolan is a natural volcanic material coming from People's Republic of Cape Verde, in powder form.

Test specimens were cast with the bar kept horizontal in the axis of the mould. Concrete compaction was carried out with internal vibrators with 35 mm diameter. The specimens were removed from the moulds not earlier than 20 h after casting, and were stored in water until the testing age of 28 days.

Steel bar (ASTM grade 400 MPa minimum yield level) with a nominal diameter of 20 mm was tested in accordance with requirements of ASTM A 615M-86a, and presented a yield force of 431 MPa. In the bond tests, the bar had 1000 mm in total length and crossed the cube from one side to the other leaving a free end of 20 mm (for slip measurement) and a 780 mm long end to receive the tensile force.

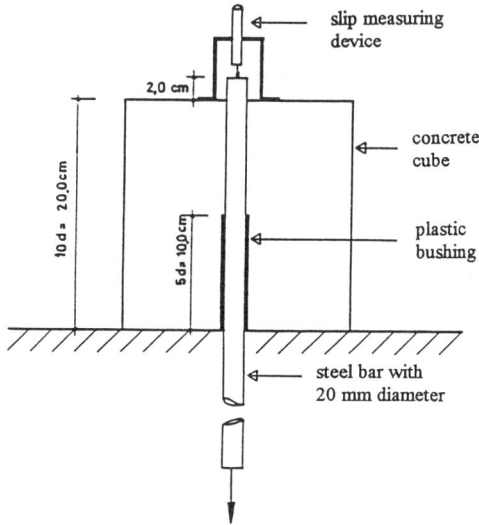

Figure 1 Schematic representation of pull-out test

The universal testing machine used for these determinations fulfilled the requirements of the relevant standards; slip of the bar relative to concrete was measured at the unloaded end of the bar, using an electrical deflectometer LVDT W10 capable of measuring slip up to ± 10 mm, with a sensitivity of 10^{-2} mm. This instrument rested on a metal support adhering to the concrete and was connected to a recorder that simultaneously plotted the testing force-slip curves. A splitting mode of failure was observed in all test specimens. Figure 2 shows the results obtained from an average of two test specimens per concrete. In this figure, on the right, the bond stress (MPa) was calculated by dividing the load on the bar (N) at any stage in the test by the nominal surface area of the embedded length of the bar in contact with the concrete (6,437 mm^2), in conformity with ASTM recommendations [8].

Simultaneously, the compressive and tensile strength of concrete at 28 days were determined. The results are also given in Table 1.

DISCUSSION

Experimental Results

As Figure 2 shows, the highest bond stresses were measured in the concrete containing superplasticizers (B3, B4, B2, B7 and B8). In this group, concrete with CSF (B3 and B4) performed better than concrete with 100% cement (B2). The contrary happened with concrete with PFA and natural pozzolan.

The concrete without superplasticizers presented more similar results, making it difficult to distinguish any specific influence of the different mineral admixtures on bond. Anyway it seems that concrete for which a part of cement was replaced by a mineral admixture (B5, B6, B9 and B10) shows slightly better bond results than the one of the reference concrete with 100% cement (B1).

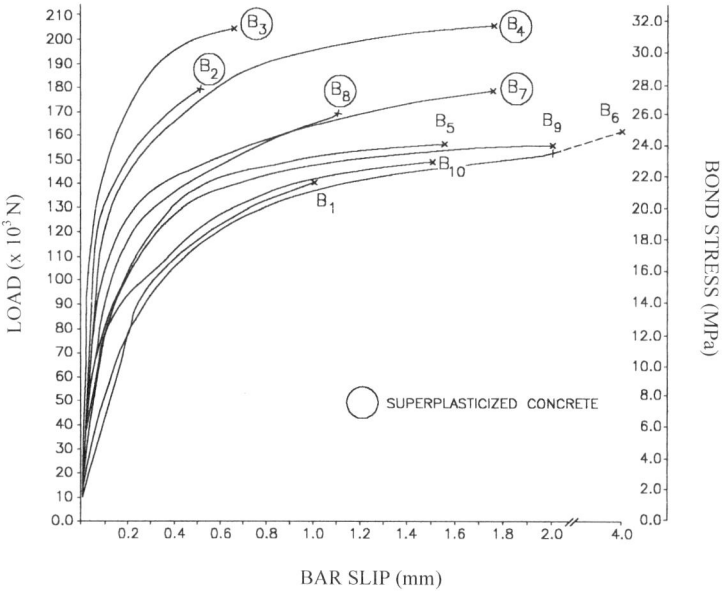

Figure 2 Force-slip curves

Figure 3 shows that bond in general varied in inverse proportion to the W/C of concrete, and that the two types of concrete with CSF A (B3 and B5) provided better results than those estimated by the general relation.

Figure 4 shows that bond varied along with the compressive strength of concrete, no particular result departing from this trend.

Comparison Between the Experimental Results and the Code Predictions

In the second part of the work, it has been decided to compare the anchorage lengths obtained from the experimental data with the ones deduced from the previsions of CEB/FIP MC 90 and of the Norwegian Standard, two international standards that were updated and are already able to handle with high strength concrete. Mention should be made of the fact that, in order to consider only the best results obtained in terms of compressive strength and bond stress, concrete B3 has been selected to be thus analyzed. In order to evaluate comparatively the performance of the admixtures in concrete, its companion concrete without superplasticizer (B5), and the 100% cement concrete (B1 and B2) were also analyzed.

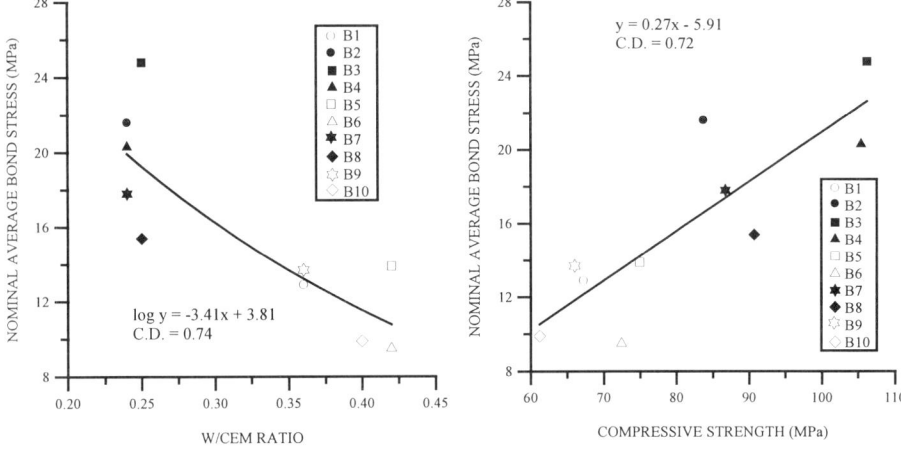

Figure 3 Correlation between bond stress and W/C of concrete

Figure 4 Correlation between bond stress and compressive strength of concrete

Theoretical Bata Based on CEB/FIP MC 90

According to this code,
$$l_b = \frac{\phi}{4} \frac{f_{syd}}{f_{bd}}$$

where:

l_b = basic anchorage length (mm)
ϕ = bar diameter (mm)
f_{syd} = design yield force of the bar ($f_{syd} = f_{syk}/\gamma_s$) (MPa)
f_{bd} = design bond stress (MPa)

The design value of the bond stress f_{bd} is: $f_{bd} = \eta_1 \cdot \eta_2 \cdot \eta_3 \cdot f_{ctd}$

where:

f_{ctd} = design value of concrete tensile strength (MPa);
η_1 = factor for the type of reinforcement, and η_1 = 2,25 for ribbed bars;
η_2 = factor for the position of the bar during concreting, and η_2 = 1,0 when good bond conditions are obtained
η_3 = factor for bar diameter, and η_3 = 1,0 for $\phi \leq 32$ mm

Table 2 shows a list of the values obtained with the application of the above formulae to the data of the four types of concrete selected.

Table 2 Values of anchorage length determined in accordance with MC 90

MIX	f_{ctd} * (MPa)	f_{bd},theor (MPa)	l_b,theor** (mm)	$\tau_{b, max}$ (MPa)	f_{bd},exp (MPa)	l_b,exp*** (mm)	DIFFERENCE (%) BETWEEN THEOR. AND EXP. VALUES
B1	2.7	6.0	312	21.5	14.3	131	138
B2	3.6	8.1	231	27.5	18.3	102	126
B3	4.5	10.2	184	31.2	20.8	90	104
B5	3.5	7.8	240	24.0	16.0	117	105

*   Design value of axial tensile strength determined in accordance with MC 90.
** Determined from the calculated value of f_{bd}
*** Determined from the experimental value of f_{bd}

Theoretical Data Based on Norwegian Standard NS 3473 E

According to this document, $l_b = 0{,}25\ \phi\ \sigma_s\ f_{bd} + t$

where:

ϕ = bar diameter (mm)
σ_s = calculated stress in the reinforcement bar in ultimate limit state at the cross section (MPa)
t = specified longitudinal tolerance for the position of the bar end. If such tolerances are not specified on the drawings the value of t shall not be taken less than 3ϕ
f_{bd} = design bond strength (MPa), and $f_{bd} = f_{bc} + f_{bs} \leq 2k_1\ f_{td}$

The portion f_{bs} that takes into account the influence of the transverse reinforcement was not considered.

$$f_{bc} = k_1\ k_2\ f_{td}\ (1/3 + 2c/3\phi)$$

k_1 = factor dependent on the type of reinforcement, and k_1 = 1,4 for ribbed bars,
k_2 = factor dependent on the reinforcement spacing, and k_2 = 1 in this case,
c = the least of the cover dimensions.

The condition $f_{bd} \leq 2k_1\ f_{td}$ can be written, in this case, as: $f_{bd} \leq 2.8\ f_{td}$

Table 3 shows a list of the values obtained from the above formulae for the data of the four types of concrete selected.

Table 3 Values of anchorage length, determined in accordance with NS 3473 E

MIX	f_{ctd} (MPa) *	$f_{bd} = 2.8\ f_{ctd}$	l_b,theor1 (mm) **	f_{bd},theor (MPa)	l_b,theor2 (mm) ***	$\tau_{b,\ max}$ (MPa)	f_{bd},exp (MPa)	l_b,exp (mm)* ***	DIFFERENCE (%) BETWEEN THEOR. AND EXP. VALUES
B1	2.7	7.6	308	12.6	208	21.5	14.3	191	61
B2	3.6	10.1	246	16.8	172	27.5	18.3	162	52
B3	4.5	12.6	209	21.0	149	31.2	20.8	150	39
B5	3.5	9.8	251	16.3	175	24.0	16.0	177	42

*   Design value of axial tensile strength determined in accordance with MC 90.
**   Determined from the calculated value of f_{bd} considering its limitation
***   Determined from the calculated value of f_{bd}
****   Determined from the experimental value of f_{bd}

From the analysis of the data showed by Tables 2 and 3, one can conclude that:

1. As should be expected since the first part of the work, concrete B3 (with superplasticizer and CSF A) is the one that requires the shortest anchorage lengths, both theoretically and experimentally.

2. The theoretical anchorage lengths calculated by the MC 90 (l_b,theor) or by the NS 3473E (l_b,theor1), when compared with the values calculated from the experimental bond stress (l_b,exp), are always higher. The Norwegian Standard expression leads to values closer to the experimental ones (50% instead of 120%, approximately).

3. Theoretical values of anchorage length, calculated by both standards were very similar (difference between 1 and 12%).

4. Anyway, the Norwegian Standard permitted, in this case, a very precise calculation of the anchorage length, as can be seen by comparing the experimental values of the anchorage length (9th column of table 3) with the theoretical ones (l_b,theor2 - 6th column of the Table 3), since the imposed limitation represented by the portion $f_{bd} \leq 2k_1\ f_{td}$ has not been considered

CONCLUSIONS

1. Cement replacement by mineral admixtures, except in the case of silica fume, resulted in a reduction of the bond of concrete to steel, compared with the mixtures with 100% cement. On the other hand, the utilization of a superplasticizer to reduce water content, substantially increased the concrete-steel bond.

2. Bond stresses varied inversely with W/C and directly with compressive strength of concrete.

3. The anchorage length can be consistently estimated by both codes analyzed, but the expressions presented by the Norwegian Standard give values closer to the experimental.

ACKNOWLEDGEMENTS

The authors would like to thank CAPES/MEC, PROPP/ /UFF and CNPq/ SCT (Brazil) for the financial support, and LNEC (Portugal) for technical support provided in carrying out this research.

REFERENCES

1. GJORV, O E, MONTEIRO, P J M, & MEHTA, P K. Effect of Condensed Silica Fume on the Steel-Concrete Bond. ACI Materials Journal Vol. 87, No. 6, 1990, pp. 573-580.

2. ROSEMBERG, A M, & GAIDIS, J, M. A New Mineral Admixture for High Strength Concrete. Concrete International, ACI, Vol. 11, No. 4, 1989, pp. 31-36.

3. ISPAS, T, & IONESCU, I. Production of Fly Ash Concretes Using Superplasticizers. ACI SP-91, 1986, Vol. 1, pp. 763-778.

4. RIXOM, M R, & MAILVAGANAM, N,P. Chemical Admixtures for Concrete. 2nd Ed., 1986, E.& F.N. Spon, London.

5. ALMEIDA, I R. Bond Between Reinforcing Steel and High Strength Concrete. BHP 96 - 4th International Symposium on Utilization of High-Strength/High-Performance Concrete, may/96, Paris (France). Proceedings, Vol. 3, pp. 1097-1104.

6. COMITÉ EURO-INTERNATIONAL DU BETON. CEB-FIP Model Code 1990. Thomas Telford, 1993, 457 pp.

7. NS 3473 E. Concrete Structures. Design Rules (1992). Norwegian Council for Building Standardization, 4th ed., 78 pp.

8. ASTM C 234-86. Standard Test Method for Comparing Concretes on the Basis of the Bond Developed with Reinforcing Steel. Annual Book of ASTM Standards, Vol. 04.02, 1987, p. 193-199.

9. RILEM DOC. 7.11.128. Essais Portant sur l'Adhérence des Armatures du Béton 2, Essais par Traction. Matériaux et Construction, Vol. 3, No. 15, 1970, pp. 175-178.

10. RILEM/CEB/FIP. Recommendations on Reinforcement Steel for Reinforced Concrete, RC 6, Bond Test for Reinforcement Steel, 2, Pull-out Test. CEB News No. 73, 1983.

EFFECTS OF DIFFERENTIAL TEMPERATURE AND SHRINKAGE ON PRE-CAST AND CAST-IN-SITU COMPOSITE CONCRETE CONSTRUCTIONS

T W Leong
National University of Singapore
Singapore

ABSTRACT. A theoretical analysis is made of the time dependent stresses and strains in precast and cast-in-situ composite concrete flexural members subjected to the effect of differential shrinkage and temperature. The behaviour of the precast portion of the member is analysed in stage 1, and that of the composite member in stage 2. These analyses have shown that substantial redistributions of stresses and strains occur over the sections at the two stage member according to time. The quantities and positions of the reinforcing steel in the section(s) contributed greatly to such redistributions. The analysis also provides a means to evaluate the potential of tensile cracking in the concrete member at any given time in stage 2, provided the tensile strength of the member is known.

Keywords: Precast and cast-in-situ composite concrete member, Differential temperature and shrinkage, Creep, Relaxation of stress, Structural analysis, Tensile strength, Time dependent stresses and strains, Cracks in concrete

Dr Tuck Wah Leong is an Associate Professor teaching Structures, Concrete Technology and Building Defect Diagnostics, at the school of Building and Real Estate, National University of Singapore. Other than teaching and research he has been working as Professional Engineer in Singapore and Malaysia and is also an Accredited Checker in Singapore for large numbers of civil and structural engineering projects.

INTRODUCTION

The development of stress and strain in precast and cast-in-situ composite concrete member under the effects of differential temperature and shrinkage is relatively complex. Even before the integration of the structural topping, constant or differential temperature and shrinkage can produce longitudinal movement in the unloaded precast concrete member. Strains and stresses are induced due to the restraints provided by the longitudinal reinforcing steel. Due to differential temperature and shrinkage, together with the provision of a preponderance of main reinforcement near one face, non-uniform stresses and strains developed over the member sections. If the temperature and shrinkage movement is acting in unison (i.e the most severe case) maximum strain occurs at a face furthest away from the restraining reinforcement and varies to the minimum value at the opposite face. The maximum stress in the concrete thus occurs in the face closer to the main reinforcing steel and vary to a minimum in the other face (i.e coincides with the face of maximum strain.). Warping of the element occurs due to the resulting force in the concrete acting eccentrically from the sectional centroid. The integration of topping concrete at a later stage has effectively changed the configuration of the section as well as the associated sectional properties. Upon hardening of the topping concrete the perfact bond composition has caused a redistribution of stresses and strains. The whole precast concrete element is now acting as a restraint against the differential movement of the composite member. Further stresses and strains developed being a result to achieve equilibrium of forces and bending moments in the overall composite section. In either of the above two stages of stress and strain developments, it is important to note that further complication exists since under sustained stress, creep occurs in concrete. Likewise, the confined deformation also causes some relaxation of the sustained stress. Creep and relaxation of stress cause an ever variations in the time dependent stresses and strains in the members. The two stage analysis of stresses and strains in the precast and the cast-in-situ composite concrete member under the effects of differential temperature and shrinkage are given in this paper. Example of numerical application of this analytical method for the confirmation of cracking in the concrete of precast and cast-in-situ concrete beam caused by severe differential temperature and shrinkage effects has also been included.

ANALYTICAL STUDY

Stage 1 – Pre-Cast Concrete Members

A rectangular precast concrete beam with section b x a (width x total depth) is considered. The beam is reinforced with both tensile and compressive reinforcement of the amounts respectively equal to pbd and p'bd where d is the effective depth and p and p' are the respective ratios of the tensile and the compressive steel contents i.e. p=As/(bd) and p'=As'/(bd) see Figure. 1a. The beam is considered free from external constraints. A longitudinal movement occurs along the beam due to the unique combination of temperature and shrinkage. The longitudinal movement is thus resisted by the steel reinforcements. Stresses and strains are built up in the concrete section. The distributions of stresses and strains across the section are probably non linear (see dotted curves in Figures. 1b, 1c) but for simplicity in the analysis these are assumed linearly distributed. The strain and stress distribution over the section at an initial state i.e. at time t=τ and at a later state, at time t≥τ, can thus be seen as the solid lines in Figures 1b, 1 c. The resulting forces in the steel and in the concrete are thus located as shown in Figure. 1d.

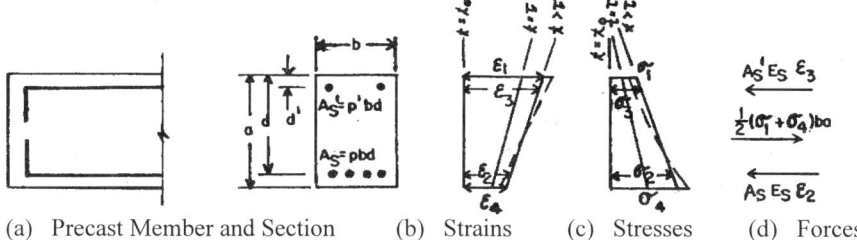

(a) Precast Member and Section (b) Strains (c) Stresses (d) Forces

Figure 1 Developments of time dependent strains and stresses due to shrinkage and temperature in precast concrete members

The equilibrium of horizontal forces and bending moment (taken with respect to the topmost beam face) can be written and simplified to the following form with d' defined as in Figure1b,

$$\sigma 1 + \sigma 4 = 2pEs\ [\{\beta(1-\beta) + \beta r(1-\delta)\}\varepsilon 1 + \beta\ (\beta + r\delta)\varepsilon 4] \text{\textemdash}(1)$$
$$\sigma 1 + 2(\sigma 4) = -6pEs\ [\beta(1-\beta) + r\delta\beta\ (1-\delta)\}\varepsilon 1 + \beta\ (\beta^2 + r\delta^2)\varepsilon 4] \text{\textemdash}(2)$$

Where $\sigma 1$, $\sigma 4$, $\varepsilon 1$, $\varepsilon 4$ are the time dependent stresses and strains in the concrete and $\beta = d/a$, $\delta = d'/a$, $r = p'/p$. In these equations the time dependent concrete strains in the levels of the reinforcements had been represented by the strains at the two extreme faces, $\varepsilon 1$ and $\varepsilon 4$, i.e. $\varepsilon 2 = \varepsilon 1(1-\beta) + \beta\varepsilon 4$, $\varepsilon 3 = \varepsilon 1(1-\delta) + \delta\varepsilon 4$

The time dependent stresses in the concrete, i.e. $\sigma 1$ and $\sigma 4$ can thus be expressed in terms of the time dependent strains, $\varepsilon 1$ and $\varepsilon 4$, i.e. from (1) (2)

$$\sigma 1 = pEs\ (Q1\ \varepsilon 1 + Q2\ \varepsilon 4) \text{\textemdash}(3)$$
$$\sigma 4 = pEs\ (Q3\ \varepsilon 1 + Q4\ \varepsilon 4) \text{\textemdash}(4)$$

where $Q1 = -2\beta[(1-\beta)\ (2-3\beta) + r(1-\delta)\ (2-3\delta)]$ $\quad Q2 = -2\beta[\beta(2-3\beta) + r\delta\ (2-3\delta)]$
$\quad\quad\ \ Q3 = 2\beta[(1-\beta)\ (1-3\beta) + r(1-\delta)\ (1-3\delta)]$ $\quad Q4 = 2\beta[B(1-3\beta) + r\delta\ (1-3\delta)]$

The development of strain in concrete must consider the occurrence of creep and the relaxation of stress in time depending on the age when the stress is sustained. The present approach follows the method proposed by Trost [1] and Bazan [2] using a creep function, ϕ, and an age adjusted modulus, X to account for these two quantities. Accordingly, the variable time dependent concrete stress caused by creep is added to its elastic component and modified by an age adjusted modulus to give the corresponding time variation in the strain. By adopting such expressions together with the shrinkage and temperature strain components, the strain expressions for $\varepsilon 1$ and $\varepsilon 4$ are,

$$\varepsilon 1 = \sigma 1\ (1 + Xo\phi)/Eo + Xo\phi\varepsilon sno/\phi no + \alpha\Delta To \text{\textemdash}(5)$$
$$\varepsilon 4 = \sigma 4\ (1 + Xo\phi)/Eo + Xo\phi\ \varepsilon sno/\phi no + \alpha\Delta To \text{\textemdash}(6)$$

Where α is the coefficient of temperature expansion (or contraction) and E is the elastic modulus of concrete. ϕno and εsno are the end creep function and shrinkage value (i.e. at time ∞); ΔT is the range of temperature considered at this stage. The symbol o represents the occurrence of event initiated from age zero e.g. E at age zero is Eo.

The creep function ϕ is normally determined from test. In the absence of test data the exponential form, $\phi = \phi n\,(1-e^{-t/tn})$ is adopted. Thus ϕ is correlated with time t in term of tn (creep year). The development of shrinkage is assumed in affinity with that of creep. From Eqs 3,4,5,6, $\varepsilon 1$ and $\varepsilon 4$ can be determined in term of ϕ, i.e. denoting n=Es/Eo

$\varepsilon 1 = [1+np\,(1+Xo\phi)]\,(Q2-Q4)]\,(Xo\phi\varepsilon sno/\phi no+\alpha\Delta To)/D$ ------------ (7)
$\varepsilon 4 = [1+np\,(1+Xo\phi)]\,(Q3-Q1)]\,(Xo\phi\varepsilon no/\phi no+\alpha\Delta To)/D$ -------------- (8)

in which $D = [1-npQ1(1+Xo\phi)]\,[1-npQ4(1+Xo\phi)]-n^2\,p^2\,Q2\,Q3\,(1+Xo\phi)^2$.

The corresponding time dependent stresses, $\sigma 1$ and $\sigma 4$ can thus be obtained from Eqs (3) (4). The time dependent steel stresses can also be obtained by substituting $\varepsilon 1$ and $\varepsilon 4$ (from Eqs 7,8) into $\varepsilon 2$ and $\varepsilon 3$ and then multiply such expression with the elastic modulus of the steel. The time dependent curvature and mid-span deflection of the precast concrete member can thus be computed, i.e. if L is the span of member,

$1/\rho = (\varepsilon 1-\varepsilon 4)/a = np\,(Q1+Q2-Q3-Q4)\,(Xo\phi\varepsilon sno/\phi no+\alpha\Delta To)/(D.a)$ -------------(9)
and $\Delta = (\varepsilon 1-\varepsilon 4)\,L^2/(8a)$ ---(10)

Stage 2 – Composite Concrete Members

At time $t=\tau$ corresponding to $\phi = \phi\tau$ (from $\phi = \phi o$), a cast-in-situ structural concrete slab of width x total depth B x T (in which B =u.b, and T=v.a, see Figure 2a) is integrated on top of the precast member. The grade of concrete used in the topping concrete can either be similar or different to that used in the manufacturing of the precast concrete. The topping slab is reinforced with steel bars of area = λ.pbd which is placed at a distance q.a from the top face of the slab (Figure. 2a). The cast-in-situ concrete normally requires some time to develop it's strength. Hence t1 is used to represent the end of stage 1 with the stresses, strains at this state denoted by the symbol t1 e.g. $\sigma 4t1$, $\varepsilon 1t1$. τ is used to represent the stage in which an assumed (perfect) bond strength had already been established in the interface and a redistribution of stresses, strains had been completed due to the change in sectional properties caused by a transformation from the stage 1, precast concrete beam, to the stage 2, composite concrete member. The stresses and strains obtained at τ are denoted by the symbol τ e.g. $\sigma 4\tau$, $\varepsilon 1\tau$. It can be seen in the later analysis that τ coincides with $\phi 2=0$ where $\phi 2$ is the creep function of the structural topping concrete. The assumption of perfect bond in the interface requires that the curvature in the cast-in-situ slab from time $t>\tau$ must equal to the curvature as that developed in the precast concrete beam. Accordingly, the strains, stresses occur in the composite beam sections caused by differential shrinkage and temperature movement must follow such distribution patterns in which the slopes of the strains and stresses occur in the slab and in the precast beam (at time $t>\tau$) must equal respectively to one another . Typical distributions of stresses and strains in the composite beam section at a time $t>\tau$ can thus be seen in Figures 2b, 2c. In this Figureures it is shown that stresses and strains do not occur in the slab until $t=\tau$ and that the time dependent stresses and strains occur in stage 2 at time t from $t=\tau$ are denoted by the script prime (') e.g. $\varepsilon 3'$, $\sigma 5'$. The stresses and strains in the precast beam can also be considered as composing of respective stresses and strains at time $t=\tau$ to that of the stage 2 components e.g. the total time dependent strain $\varepsilon 1$ at time $t>\tau = \varepsilon 1\tau+\varepsilon 1'$ and the time dependent concrete stress at level 4 i.e.$\sigma 4=\sigma 4\tau+\sigma 4'$.

The resultant (horizontal) forces in the concrete and in the steel are thus obtained as shown in Figure. 2d.

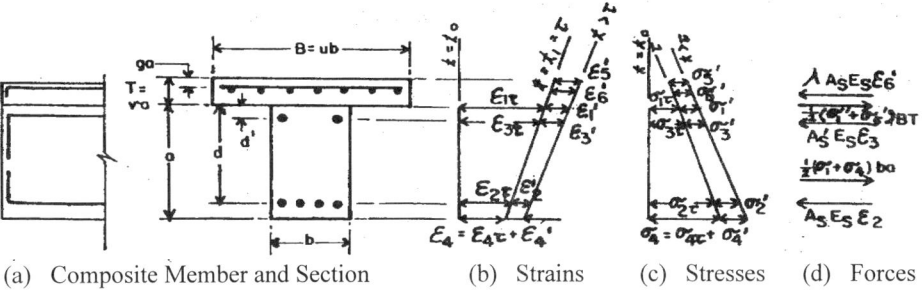

(a) Composite Member and Section   (b) Strains   (c) Stresses   (d) Forces

Figure 2 Developments of time dependent strains and stresses due to shrinkage and temperature in the composite member

The equilibrium of horizontal forces and bending moments(taking with respect to level 5) are,

$\sigma 1\tau + 2pEs(\varepsilon 2\tau + r\varepsilon 3\tau) + \sigma 1'(1+uv) + \sigma 4 + uv\sigma 5' + 2pEs(\varepsilon 2' + r\varepsilon 3' + \lambda \varepsilon 6') = 0$ ----------------------(11)

$(3v+1) \sigma 1\tau + 6Esp[r(v+\delta)\varepsilon 3\tau + (v+\beta)\varepsilon 2\tau] + \sigma 1'(3v+1+2uv^2) + uv^2 \sigma 5'$
$+(3v+2)\sigma 4 + 6Esp[\lambda q\varepsilon 6' + r(v+\delta)\varepsilon 3' + (v+\beta)\varepsilon 2'] = 0$ --------------------------------------- (12)

The non-uniform but linear distribution of the second stage time dependent strains allow the concrete strains $\varepsilon 1'$, $\varepsilon 2'$, $\varepsilon 3'$ and $\varepsilon 6'$ to be represented by those occurring at the two extreme faces, i.e. $\varepsilon 4'$ and $\varepsilon 5'$,

$\varepsilon 1' = (v\varepsilon 4' + \varepsilon 5')/(1+v)$ ---------------- (13) $\varepsilon 2' = [(v+\beta)\varepsilon 4' + (1-\beta)\varepsilon 5']/(1+v)$ ------- (14)
$\varepsilon 3' = [(v+\delta)\varepsilon 4' + (1-\delta)\varepsilon 5')]/(1+v)$ ----- (15) $\varepsilon 6' = [q\varepsilon 4' + (1+v-q) \varepsilon 5']/(1+v)$ ---------- (16)

The time dependent relationship for the stress strain development in the second stage over the cross section of the composite beam can be obtained by adopting the same types of expressions given in Equations 5,6. Using $\phi 2$, X2, $\phi n2$, $\varepsilon sn2$ to represent the creep function, the age adjusted Modulus (or relaxation coeff.), the end creep function and the end shrinkage values for the concrete in the slab. The strains and stresses developed at levels 1, 5 (Figures.2b, 2c) are,

$\varepsilon 1' = \sigma 1' (1+X2\phi 2)/E\tau + X2\phi 2 \; \varepsilon sn2/\phi n2 + \alpha 2 \; \Delta T2$ ------------ (17)
$\varepsilon 5' = \sigma 5' (1+X 2\phi 2)/E\tau + X2\phi 2 \; \varepsilon sn2/\phi n2 + \alpha 2 \; \Delta T2$ ------------- (18)

However, at the soffit of the precast beam the developments of shrinkage, temperature and creep strains have different values from those in the slab levels, i.e. from Equation 6,

$\varepsilon 4 = \varepsilon 4\tau + \varepsilon 4' = (\sigma 4\tau + \sigma 4')[1 + Xo(\phi\tau + \phi 2)]/Eo + Xo(\phi\tau + \phi 2) \; \varepsilon sno/\phi no + \alpha \Delta To$ ------------------(19)

in which $\varepsilon 4\tau$, $\sigma 4\tau$, $\phi\tau$, Xo, ϕno, εsno, Eo have all been defined earlier and are qualities related to the stage1 development of stresses and strains in the precast member.

The time dependent stresses, $\sigma1'$, $\sigma4$ (or $\sigma4\tau+\sigma4'$), and $\sigma5'$ in Eqs 11,12 can thus be expressed in terms of $\epsilon4'$ and $\epsilon5'$, i.e. by equating Eqs (13) to (17) and from Eqs (18), (19) respectively,

$\sigma1'=E\tau[(\nu\epsilon4'+\epsilon5')-(1+\nu)(X2\phi2\epsilon sn2/\phi n2+\alpha2\Delta T2)]/[1+\nu)(1+X2\phi2)]$ -------------------------(20)
$\sigma5'=E\tau[\epsilon5'- X2\phi2\epsilon sn2/\phi n2-\alpha2\Delta T2]/(1+X2\phi2)$ ---(21)
$\sigma4=\sigma4\tau+\sigma4'=Eo[(\epsilon4\tau+\epsilon4')- Xo(\phi\tau+\phi2)\ \epsilon sno/\phi no\ -\alpha\Delta To]/[1+Xo(\phi\tau+\phi2)]$ ------------------ (22)

By substituting Equations 13 to 22 into Equations 11,12 and after simplifications, two equations governing the time variations in strains at level 4 and 5 respectively are thus obtained,

D1 $\epsilon4'$+D4 $\epsilon5'$ = L2 N11- R1 L2 Ft -- (23)
D3 $\epsilon4'$+D2 $\epsilon5'$ = L2 N13- R2 L2 Ft -- (24)

and by solving for $\epsilon4'$ and $\epsilon5'$

$\epsilon4'$ = L2 {N11 D2-N13 D4-L2Ft (R1D2-R2 D4)}/D --- (25)
$\epsilon5'$ = -L2 { N11D3-N13D1-L2Ft (R1D3-R2D1)}/D --- (26)

where D = D1D2- D3 D4   and   D1 = L1Ftτ+L2+S1Ft   D2 = L7 Ftτ+S4 Ft
D3 = L5 Ftτ+L6 L2+S3 Ft   D4 = L3 Ftτ+S2 Ft   and
N11 = L4 Ftτ (X2ϕ2ϵsn2/ϕn2+α2ΔT2)+ Xo($\phi\tau$+ϕ2) ϵsno/ϕno+$\alpha\Delta$To-ϵ4τ
N13 = L8 Ftτ (X2ϕ2ϵsn2/ϕn2+α2ΔT2)+ L6{ Xo($\phi\tau$+ϕ2) ϵsno/ϕno+$\alpha\Delta$To-ϵ4τ}

The modulus of elasticity for the old and the new concrete can be different. The ratio Eτ/Eo is thus defined as equate to 1/g where g is considered a constant. The above expressions contain not only constants but also variables with time t, τ as the parameters. The constants are,

L1=ν(1+uν)/g;   L2=1+ν;   L3=(1+2uν+uν^2)/g;
L4=(1+2uν)/g;   L5=ν(1+3ν+2uν^3)/g;   L6=3ν+2;
L7=(1+3ν+3uν^2+ uν^3)/g;   L8=(1+3ν+3uν^2)/g;

R1=($\sigma1\tau+\sigma4\tau$)/Eo+2np ($\epsilon2\tau$+r$\epsilon3\tau$) whereas at time t1, ϵ2t1, ϵ3t1 shall be used in R1 and R2,
R2=[(3ν+1) $\sigma1\tau$+(3ν+2) $\sigma4\tau$]/Eo+6np[r(ν+δ) $\epsilon3\tau$ + (ν+β) $\epsilon2\tau$]
S1=2np{(ν+β)+ r(ν+δ)+λq}   S2=2np{(1-β)+ r(1-δ)+ λ(1+ν-q)}
S3=6np{λq^2+r(ν+δ)^2+(ν+β)^2]   S4=6np [λq(1+ν-q)+ r(ν+δ) (1-δ)+(ν+β) (1-β)]
and the variables are Fτ = 1+X2ϕ2   Ft=1+ Xo($\phi\tau$ + ϕ2)   Ftτ = Ft /Fτ

Equations 25,26 allow the time varying strains at stage 2 to be evaluated according to time in terms of ϕ2 which can be linked to the real time scale using the creep function and time correlation. With $\epsilon4'$ and $\epsilon5'$ computed according to a given ϕ2 value, the other time varying stresses and strains over the composite section can thus be determined from Equations 13 to 16 and 20 to 22.

The total stresses in level 1 at time corresponding to a creep function ϕ2 or ϕ = $\phi\tau$+ϕ2 is thus $\sigma1'$ (obtained from Eq20, occurring in soffit of slab) and $\sigma1$ = $\sigma1\tau$ +$\sigma1'$(occurring on top of the precast beam).

REDISTRIBUTION OF STRESSES AND STRAINS

The strains and stresses in the composite section at time $t=\tau$ can be obtained by substituting $\phi 2 = 0$, $\Delta T2 = 0$ into Equations 23,24 to obtain $\varepsilon 4'$ and $\varepsilon 5'$ which are in turn added to $\varepsilon 4t1$, $\varepsilon 5t1(=0)$ and denoted as $\varepsilon 4\tau$ and $\varepsilon 5\tau$. By substituting these two values into Equations 13 to 16 and 20 to 22, $\varepsilon 2'$, $\varepsilon 3'$, $\varepsilon 1'$, $\sigma 1'$, $\sigma 5'$ and $\sigma 4'$ can thus be determined. In this process, however, the existing stresses and strains, i.e. $\varepsilon 2t1$, $\varepsilon 3t1, \varepsilon 1t1$, $\sigma 1t1$, $\sigma 4t1$ shall be added to the $\varepsilon 2'$, $\varepsilon 3'$, $\varepsilon 1'$, $\sigma 1'$, and $\sigma 4'$ respectively to give $\varepsilon 2\tau$, $\varepsilon 3\tau$, $\varepsilon 1\tau$, $\sigma 1\tau$, $\sigma 4\tau$ which are the initial stresses and strains corresponding to $\phi 2 = 0$.

BEHAVIOUR UNDER DIFFERENTIAL SHRINKAGE AND TEMPERATURE

Two sets of data are chosen to illustrate the variations in stresses and strains in a precast concrete beam in stage 1. The beam is further incorporated with a structural topping slab in stage 2. The precast beam has dimensions bxa = 250mmx500mm and the structural topping concrete, has dimensions BxT = 1350mmx100mm. Other physical, sectional and mechanical properties for the precast and then made composite beam are:

In stage 1:

$\beta=0.9$ $\delta=0.08$ $\alpha o = 11 \times 10^{-6}/°C$, $Xo = 0.67$, $\phi no=1.5$, $\varepsilon sno=0.00015$, $p=0.03$, $r= 0.5$, $\Delta To=5°C$, $Eo = 20$ KN/sqmm

In stage 2

$v=0.2$ $u=5.4$, $\lambda=0.05$, $q=0.06$, $\alpha 2 = 11 \times 10^{-6}/°C$, $X2 = 0.87$, $\phi n2 =3.0$, $\varepsilon sn = 0.0003$, $\Delta T2 =20$ °C, $E\tau=15$KN/sqmm. The Modulus of elasticity for the steel bars is 200KN/sqmm.

The two sets of data (example 1 and 2) are varied in the ages in which stage 1 is respectively ended, i.e. $\tau = 0.15\phi no$ and $\tau=0.75\phi no$. Relatively small values of ϕno, εsno and ΔTo are selected for the precast beam at stage 1 and compare with those correspondingly greater values for the composite beam at stage 2 emphasizes the characteristic of this type of problems. The analytical results are summarized in Tables 1, 2. It can be seen that the integration of a structural topping at end of stage 1 i.e. at the corresponding t1 or τ value causes redistribution of strains and stresses from the precast member to the composite member. The redistribution of these strains and stresses are relatively small before stage 2 differential shrinkage and temperature have been incorporated. Although differential temperature and shrinkage have produced great variations in the time dependent stresses and strains, such effects are now taken up by a relatively stiffer composite beam.

The severe effects have been relatively 'defused' to some extent which can be witnessed in the more even out numerical figures between concrete stresses $\sigma 1$ and $\sigma 4$ at various states of development for $t >\tau$. A even more uniform distribution of time dependent stresses in both, stages 1 and 2 is observed in precast beams containing equal amounts of tension and compression steel at approximately symmetrical positions. This can be seen from the results of analysis (Example3) on a set of data containing numerical values of parameters basically similar to those given in the two above examples except that $p=0.01$, $r=1.0$ $\Delta To=10°C$, $\varepsilon sno=0.003$, $\phi no=3.0$, $\Delta T2 =15°C$, $\varepsilon sn2=0.00035$, $\phi n =3.5$.

Table 1 Result of analysis (Example 1) $\phi\tau=0.15\phi no$, stresses in N/mm^2, strain in 10^{-6}

STAGE	ϕ or $\phi\tau$	$\epsilon 1'$	$\epsilon 1$	$\epsilon 4$	$\epsilon 5'$	$\sigma 1'$	$\sigma 1$	$\sigma 4$	$\sigma 5'$
1	$\phi = \phi o = 0$	-	55	33	-	-	-0.004	-0.43	-
	$\phi = \phi 1 t = 0.15$	-	65	38	-	-	-0.001	-0.49	-
2	$\phi t = \phi\tau = 0.15$	-2	63	46.5	-4	-0.027	-0.028	-0.34	-0.057
	$\phi t = 0.75$	213	276	48	255	-0.584	-0.560	-0.76	-0.168
	$\phi t = 1.25$	237	300	55	283	-0.600	-0.630	-0.92	-0.253
	2.15	279	342	62	331	-0.630	-0.660	-1.12	-0.344
	3.15	313*	376*	49*	375*	-.700*	-.728*	-1.0*	-0.440*

Results affected by ϵso reaching ϵsno, the limiting value

Table 2 Result of analysis (Example 2) $\phi\tau=0.75\phi no$, stresses in N/mm^2, strain in 10^{-6}

STAGE	ϕ or $\phi\tau$	$\epsilon 1'$	$\epsilon 1$	$\epsilon 4$	$\epsilon 5'$	$\sigma 1'$	$\sigma 1$	$\sigma 4$	$\sigma 5'$
1	$\phi = \phi o = 0$	-	55	33	-	-	-0.004	-0.43	-
	$\phi = 0.50$	-	87	48	-	-	-0.023	-061	-
	$\phi = 0.75$	-	102	54	-	-	-0.042	-.068	-
2	$\phi t = \phi\tau = 0.75$	-3	99	67	-6.0	-0.045	-0.087	-0.51	-0.09
	$\phi t = 1.25$	206	305	63	248	-0.60	-0.69	-0.83	-0.16
	1.75	230	329	67	276	-0.62	-0.70	-0.96	-0.25
	2.75	272*	371*	63*	326*	-0.67*	-.076*	-.099*	-.037*
	3.75	306*	405*	51*	370*	-0.73*	-0.82*	-0.88*	-0.46*

Results affected by ϵso reaching ϵsno, the limiting value

Table 3 Result of analysis (Example 3) $\phi\tau=0.5$, stresses in N/mm^2, strain in 10^{-6}

STAGE	ϕ or ϕt	$\epsilon 1'$	$\epsilon 1$	$\epsilon 4$	$\epsilon 5'$	$\sigma 1'$	$\sigma 1$	$\sigma 4$	$\sigma 5'$
1	$\phi = \phi o = 0$	-	88	90	-	-	-0.33	-0.31	-
	$\phi = \phi_1 t = 0.5$	-	107	109	-	-	-0.41	-037	-
2	$\phi t = \phi\tau = 0.5$	2	109	128	-2	0.02	-0.39	-.017	-0.026
	1.0	173	281	130	207	-0.32	-0.71	-0.43	-0.014
	1.5	204	313	141	242	-034	-0.73	-0.52	-0.070
	2.5	262	370	159	308	-0.37	-0.76	-0.67	-0.150
	3.5	314	422	171	368	-0.21	-.077	-0.78	-0.210
	4.0	341	449	176	399	-0.43	-0.80	-0.82	-0.240

CRACKING DUE TO DIFFERENTIAL TEMPERATURE AND SHRINKAGE

Equations 20 or 22 can be used in a reverse process to obtain the differentiation in temperature or shrinkage occurs in between the precast and the composite members which can cause tensile cracking at either the soffit level of the topping slab or at the soffit level of the precast beam. Basically, there are two stages of shrinkage and temperature values for the precast and composite beam, i.e in terms of εsno, ΔTo at stage 1 and $\varepsilon sn2$ and $\Delta T2$ at stage 2. The end shrinkage and temperature values at stage 1 are usually considered known values since the performance of the composite member is interested herewith. If it is further assumed that $\varepsilon sn2$ is known, then $\Delta T2$ is the only unknown provided that the tensile strength of concrete for the member σt is given (in replacement of $\sigma 1'$ values at the various states in stage 2). The prediction of cracking in the concrete can be seen from the example given below.

A precast and cast-in-situ composite beam with sectional dimensions and properties at stage 1 similar to those used in Example 1 is considered herewith. In stage 2 a structure topping slab of B x T = 1200mm x 100mm is integrated at age $\phi\tau = \phi t = 1.0$. Other sectional, physical and mechanical properties are, $v=0.2$, $u=4.8$, $\lambda=0.10$, $q=0.16$, $X2=0.87$, $\phi n2=2.0$ $\varepsilon sn2=0.00035$, $E\tau=13KN/sqmm$, $Es=200KN/sqmm$ and $\alpha 2=11 \times 10^{-6}/°C$. The time dependent stresses and strains at stage 1 (occurs in the precast sections only) are evaluated in a similar manner as described in the earlier example. In stage 2, the differential temperatures, T2 (or $\Delta T2$) which will cause cracking in the soffit of the cast-in-situ slab are considered unknown at four different time represented by $\phi 2$ or $\phi t= \phi\tau + \phi 2$ where $\phi 2$ have been assigned with the values 0 (i.e. $\phi 2=\phi\tau$), 0.5 ($\phi t= \phi\tau + 0.5$), 1.0 and 1.5. The time dependent strains $\varepsilon 4'$, $\varepsilon 5'$ at these various ϕt values can thus be obtained from Equations 25,26 with $\Delta T2$ as the only unknown. By substituting the corresponding $\varepsilon 4'$ and $\varepsilon 5'$ (of the same state of time, ϕt) into Eq 20, an equation is obtained relating the time dependent concrete stress at the soffit of the cast-in-situ topping slab i.e. $\sigma 1'$ to $\Delta T2$, the unknown temperature which will cause cracking in the bottom flange. The following $\Delta T2$ values are obtained by further equating $\sigma 1'$ values with σt values at the respective time (i.e. ϕt):

Table 4 Result of example 4   ($\phi\tau=1.0$, To = 5 °C)

$\phi 2$	0	0.5	1.0	1.5
ϕt	1.0	1.5	2.0	2.5
σt (N/mm^2)	-0.8	-1.0	-1.2	-1.36
$\Delta T2$ (°C)	23.8	25	29	31 *
$\Delta T2 - T0$ (°C)	18.8 C	20 C	24 C	26 C

* Results affected by εso reaching εsno, the limiting value

CONCLUSIONS

The integration of a cast-in-situ structural topping to a precast beam has virtually transformed the sectional properties and stiffness of the precast section to those of the composite section through such integration the enhanced (composite) section, posessed greater capacity for redistribution of time dependent stresses and strains caused by longitudinal movements due to differential temperature and shrinkage.

The quantity and location of steel reinforcement in the precast and composite beam sections contributed significantly to the redistributions of these stresses and strains at both stages of the analysis. The strains and stresses built up in the precast beam at the end of stage 1 influenced further development of stresses and strains in the second stage. Tensile cracking in the precast and composite section, caused by differential temperature and shrinkage effectsc can be predicted, with the extent of cracking corresponding to the tensile strength of the concrete at any state in stage 2.

REFERENCES

1. TROST, H., "Answirkungen des superposition's princips auf Kriech-und Relxations – Probleme bei", Beton and stahlbetonbau Vol. 62, No.10, p230-238; No 11, p261-269, 1967.

2. BAZANT, Z.P, "Prediction of concrete creep Effects using Age-Adjusted Effective Modulus Method", ACI Journal Proc., Vol 69, No. 4 (April) p212-217 1972.

THE EFFECT OF AGGREGATES- CEMENT INTERACTIONS ON THE STRENGTH OF CEMENTTIOUS MATERIAL

G J Buckles
CERAM Research

B R Heywood   K Kendall

K Milnes
Keele University
United Kingdom

ABSTRACT. This study highlights the importance of the chemical and physical properties of concrete/mortar components in determining ultimately the strength of the final material. A range of analytical techniques (including ICP, FTIR, XRD, SEM) have been utilised to map the chemical and physical attributes of the primary components in addition to the composition and microstructure of the resulting concretes and mortars. These results have been correlated with the materials compressive strength data This study reveals that increased aggregate porosity and increased dissolution of calcium on hydration causes a decrease in the relative proportions of amorphous to crystalline hydration products formed in the resulting concrete. In addition, ionic dissolution from aggregate particles is shown to be highly influential on the hydration processes of cements. These aggregate parameters cause an alteration in the hydration sequence followed by the cement. Resulting changes that occur in the concentration and placement of $Ca(OH)_2$ in these concretes are the cause of statistically significant strength differences.

Keywords: Aggregate, Portland cement (PC), Composition, Microstructure, Characterisation.

Dr Gavin Buckles at CERAM Research is a project manager for the Refractories and Industrial Ceramics Division. Particular interests include polymer-modified cements.

Professor Brigid R Heywood is Chair of Inorganic Chemistry at Keele University. She is Research Director of the Crystal Science Group at Keele, which specialises in crystal engineering of inorganics and the microstructure/property function relationships of solid state materials.

Professor Kevin Kendall is director of the Birchall Centre for Inorganic and Materials Science at Keele University. He specialises in the chemistry of particulate materials such as polymer latex, zirconium oxide and hydraulic cements.

Kathryn Milnes is a PhD student on the PTP scheme at Keele University/CERAM Research. Her PhD is focusing on variation of microstructure in the interfacial transition zone (ITZ) of concrete and mortar.

INTRODUCTION

It is widely acknowledged that alterations in the mix components of concrete/mortar will change the mechanical properties of the resulting material, [1, 2]. Variations in the components can alter the hydration products formed either directly or indirectly, [3, 4], e.g. affecting the amount and identity of hydrated material or the morphology of the hydration products. It is clear that a greater understanding of the way in which microstructure and strength are affected by the mix composition will enhance our ability to predict the influence of small variations in component characteristics upon the materials performance. Here the effects of changing one or both of the primary components (Portland Cement, gravel aggregate) on the structure/property/function relationships of concrete have been investigated.

Background

Accumulated performance data for concretes and mortars fabricated using three Portland cements (PC-A, PC-B, PC-C) and three distinct aggregates (A-1, A-2, SS-3) of different origin but identical specifications showed significant variations in their physical performance characteristics, (see Figure 1). Statistical analyses of strength testing data showed that concretes produced with A-1 possess a lower compressive strength than concretes produced with A-2. It can also be seen, however, that when concrete is produced with A-2 there is a significant difference between the strengths of samples of this concrete when produced with PC-A and PC-C.

The origin of these strength differences has been investigated. A range of analytical techniques was utilised to map in detail the chemical and physical attributes of the primary components and the microstructure of the paste-aggregate interfacial region of the ensuing concrete. These data were then correlated with the materials compressive strength.

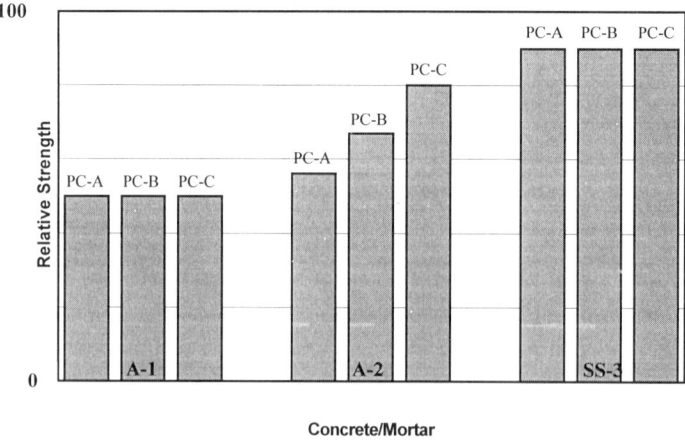

Figure 1 Plot of average concrete/mortar strengths

EXPERIMENTAL PROGRAMME

Samples of concrete and mortar were produced according to the British Standard test procedure; (BS 1881 for testing concrete and BS EN 196 for mortar testing, see Appendix 1). The samples were stored in a humidity cabinet at 20+/-1°C and +90% relative humidity for 24 hours. At an age of 24+/-4 hours the samples were demoulded and cured in lime saturated water at 20+/-1°C until the desired age. In the present study all samples were analysed at an age of 28 days. Samples of cement paste produced for analyses also followed this curing regime. At 28 days old the samples were removed from curing and specimens cut from the cubes/prisms. These specimens were dried by soaking in acetone and placed in an oven at 60°C. They were subsequently allowed to cool and stored in dessicators.

Strength data used in this investigation was collected from the compressive strength testing of concrete cubes and mortar prisms in accordance with BS 1881 for concrete and BS EN 196 for mortar samples (see Appendix 2). Fourier transform infra-red, X-ray diffraction, X-ray fluorescence and thermal analyses were carried out on powdered samples of material using respectively: a Nicolet Impact 400 FTIR; Seimens D5005 XRD; Philips PW 1606 XRF; Linseis Al81 combined TG and DTA. SEM analyses were carried out on both polished and fractured specimens using a Hitachi FESEM.

RESULTS AND DISCUSSION

Analyses of Primary Materials

The raw materials were characterised with the aim of profiling those physical and chemical properties which could influence the concrete/mortar strength both indirectly, through the paste-aggregate bond, and directly.

The three aggregates comprised two gravel aggregates and a standard sand. The two gravel aggregates (A-1 and A-2) had the same mean particle size and particle size distribution. When used in the mix these were split into three fractions - fine, 10-5mm and 20-10mm. (The standard sand (SS-3) was equivalent to the fine fraction of these). A-1 and A-2 had different water absorption values, (see Figure 2). A-1 had a water absorption profile almost twice that of A-2. These results were confirmed by mercury porosimetry.

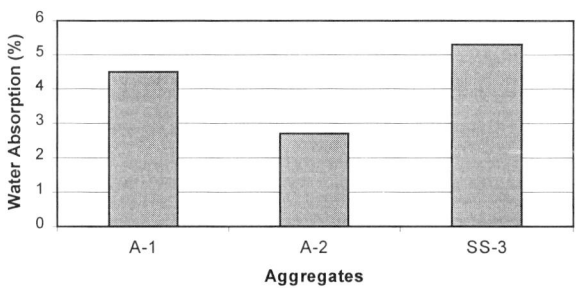

Figure 2 Water absorption profiles for A-1, A-2 and SS-3

The two aggregates also had different elemental compositions. The elemental composition, as identified by XRF, (see Figure 3), revealed that A-2 and SS-3 have very similar compositions.

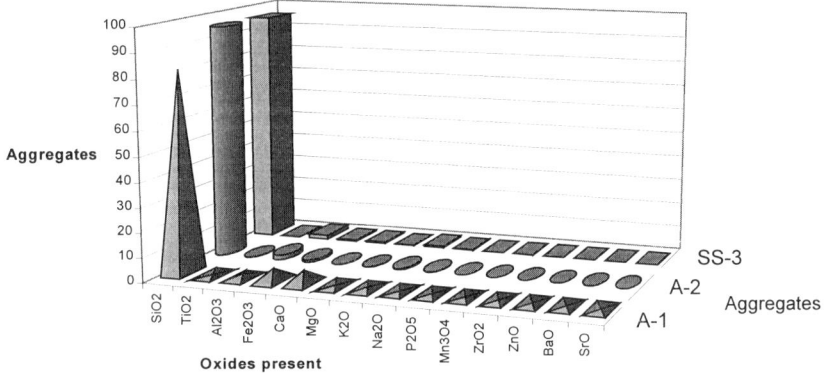

Figure 3 Results of XRF analyses of A-1, A-2 and SS-3

Both A-2 and SS-3 were composed primarily of SiO_2 (95%); A-1 on the other hand contains less SiO_2 and more CaO and Fe_2O_3 than the other aggregates, (SiO_2 = 81%, CaO = 5%, Fe_2O_3 = 4%). This again was reflected in the phase compositions identified by XRD analyses. The pH profiles of the three aggregates when mixed with water over a seven-hour period were determined. Both A-2 and SS-3 were associated with slightly acidic pH's and A-1 had a slightly basic pH. ICP analyses showed that the ions entering solution from the aggregates were simply small amounts of silica from A-2 and SS-3 and from A-1 some calcium in addition to the silica. Both the pH and ICP results reflect the compositional differences between the aggregates. The main differences between the three Portland Cements used in this investigation were in their mean particle size and specific surface area (SSA), (see Figure 4). PC-C had the smallest mean particle size and, therefore, largest SSA; PC-B had the largest particle size and smallest SSA.

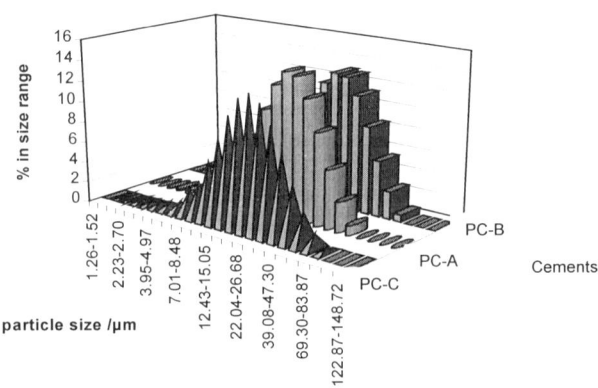

Figure 4 Results of particle size evaluations on PC-A, PC-B and PC-C

PC-C is a sulphate resisting Portland Cement and, therefore, contains no C_3A but double the amount of C_4AF than the other PC's, (see Figure 5). PC-C contains more Fe_2O_3 and less Al_2O_3 than the other PC's. In all other respects the three PC's have very similar Bogue and elemental compositions. Other analyses (FTIR, XRD and thermal) support these data, showing the only chemical difference to be in the C_4AF and C_3A phases.

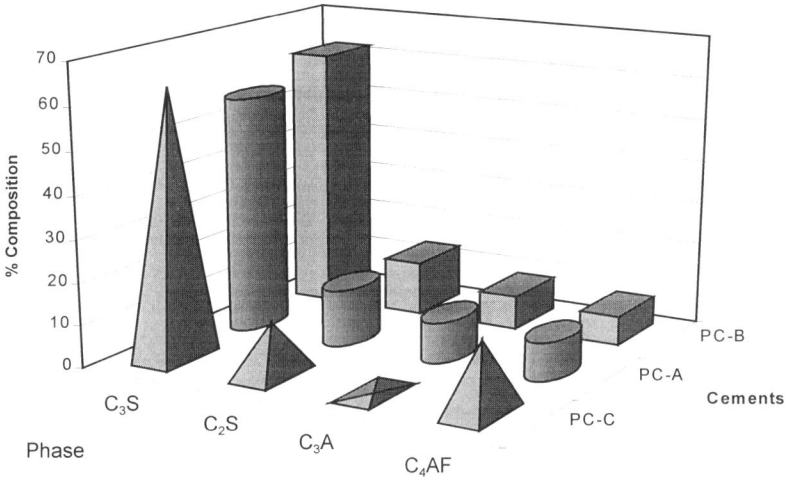

Figure 5 Bogue Composition of Cements

Samples of concrete, mortar and cement paste were produced and their composition and microstructure analysed.

Effect of Aggregate Variability on Composition and Microstructure

Concrete and mortar samples were analysed with the aim of investigating the significant differences in strength between samples produced with A-1 and A-2, (see Figure 1). Samples were produced with A-1, A-2 and SS-3 using PC-A. The standard sand was used to produce a mortar to act as a control.

FTIR, XRD and thermal analyses showed that these samples contain the same hydration products after a 28-day hydration period, (see Table 1). However, a variation in the amount of crystalline $Ca(OH)_2$ between the samples was noted; analyses of samples containing A-1 revealed more prominent $Ca(OH)_2$ than for the samples containing A-2 or SS-3. This suggested an increase in the proportion of crystalline to amorphous hydration products for concretes using A-1. SEM analyses of these materials revealed a spread of small $Ca(OH)_2$ crystals throughout samples containing A-1. In contrast samples produced with A-2 and SS-3 showed that the $Ca(OH)_2$ crystals were mainly confined to pores and the interfacial region. Also, the crystals were generally larger than those found in A-1 samples.

Table 1 Schematic summary detailing the effects of differences in the aggregates on the composition/microstructure of the concretes and mortars

AGGREGATE COMP.	POROSITY OF AGGREGATE	28DAY STRENGTH OF	CRYSTALLINITY* (FROM DTA, XRD, FTIR AND SEM)
Si Ca Fe	A-1 (high porosity)	(low)	(more crystalline)
Si Ca Fe	A-2 (medium porosity)	(medium)	(medium)
Si Ca Fe	SS-3 (low porosity)	(high)	(less crystalline)

* Crystallinity is expressed as relative amounts of crystalline (hexagons) to amorphous (shaded triangle) hydration products

Effect of Cement Variability on Composition and Microstructure

Samples of cement pastes were produced for use as control samples. These results were also used to investigate the significant differences in cement strengths induced when A-2 is used, (see Figure 1). FTIR, XRD and thermal analyses were carried out on samples of cement paste hydrated for 28 days.

Results of these analyses show that there appears to be no difference between the samples produced with different cements. XRD shows that PC-C samples contain more $Ca_2(Al.Fe)_2O_5$ (Brownmillerite) indicating the presence of more Fe_2O_3 in these samples. Hydrated PC-C cements have a slightly darker colour, reflecting the colour difference seen in the unhydrated samples. This again is most likely due to the increased amount of Fe_2O_3 in these samples. From the data collected so far for the cement samples the three cements hydrate in a similar way producing pastes that are compositionally the same.

Analyses were then carried out on mortar and concrete samples, with the aim of investigating the significant differences occurring in the ranking of the cements as a function of the anhydrous cement. Mortar and concrete samples produced with the same aggregate and different PC's were analysed.

Analyses of strength data for mortar samples produced with SS-3 and concrete produced with A-1 using PC's -A, -B and -C show no strength differences between the samples, (Figure 1). Compositional and microstructural analyses confirm these findings, revealing no differences in the bulk material between samples.

From analyses of concrete produced using A-2 and PC-A, -B and -C no real compositional differences could be identified suggesting that these materials have the same composition. Yet statistical analyses of strength data for concretes produced with A-2 show significant differences between the results when different cements are used, (see Figure 1). ICP analyses were carried out on mixtures of A-2 and the cements, determining the ions released on hydration. (See Table 2). The concentrations and relative proportions of ions released on hydration of the mix are different to those from the hydration of the components individually. This indicates a difference in the reaction sequence and/or the formation of different hydration products in the interfacial region. The three concrete systems (A2+PC's -A, -B, and -C) follow the same basic dissolution processes, although the concentrations of ions dissolved vary from system to system. These concentration differences highlight the possible evolution of chemical differences in the interfacial region of each sample.

Table 2 Results of ICP analyses on cements, aggregates and concretes produced with A-2

ELEMENTS (MMOL/G MATERIAL)	PC-A	PC-B	PC-C	A-2	A-2 + PC-A	A-2 + PC-B	A-2 + PC-C
Al	0	0	0	0.032	0	0	0
Ca	0.00064	0.0013	0.0055	5.7318	0.00021	0.00059	0.0013
K	0.014	0.075	0.031	0.027	0.0025	0.051	0.057
Na	0.00039	0	9.3E-05	0.14	0.00011	0.007	0.0068
S	0.0053	0.024	0.0083	0.64	0.00059	0.0069	0.0089
Si	1.9E-05	0.00012	8.8E-05	0.0064	0	0.00017	0.00026

It can be seen that when the mix is hydrated the concentrations of ions are different than when the components are hydrated individually. This would indicate a difference in the hydration sequelae followed and/or the formation of different relative amounts of the hydration products.

CONCLUSIONS

The microstructural and compositional differences identified in concretes produced with A-1 and A-2 are a result of the differences in the chemical and physical characteristics of the aggregates. A-1 was much more porous, (see Figure 2), and contained much more CaO and Fe_2O_3 than A-2. Porosity of aggregates is known to play a major role in determining concrete microstructure and strength, [2]. Leaching of ions from aggregates is also known to affect microstructure formation in concrete, which could in turn affect the materials strength, [5, 6]. Therefore, both of these variables in the aggregates could cause the compositional, microstructural and strength differences identified; the individual contribution of each variable has yet to be established.

The final part of this study has revealed no major differences between the microstructures of the concretes produced with A-2. This is also the case for samples of paste produced with these cements. However, an investigation of the hydration processes show that the three cements follow different hydration paths. This appears to be linked to the breakdown of phases in each particle size range on milling. It can also be seen that the presence of A-2 alters the hydration process of the cements. This alteration is probably due to a specific effect of the aggregate on the dissolution of calcium, potassium, sodium and sulphur from each cement.

APPENDIX 1

Production of Concrete Cubes According to BS 1881 Parts 102 and 108.

Aggregates and sand, oven dried to 105 +/- 5°C and cooled to 20 +/- 2°C, were weighed into a dampened mixer bowl. Anhydrous cement was weighed out and added to the mixer, the mixer and timer were started and 1000g distilled water added. The mixer was stopped after 30 seconds and left to stand for 4 minutes 30 seconds. During this time 400g distilled water were weighed into a container. Mixing was resumed for a further 3 minutes and water from the 400g added to achieve approximately a slump of 60-75mm. The mixture was left to stand for 6 minutes during which time the slump was tested. If the slump was below 60mm more water was added and the material remixed for 1 minute before retesing the slump. After determination of the slump the mixture was returned to the bowl and remixed for 15 seconds. Clean and pre-greased Tonitechnik moulds were filled in two layers, each layer was compacted with 35 tamps, distributed uniformly over the cross section of the mould.

Production of Mortar Prisms According to BS EN 196 Part 1

1350 +/- 5g sand was weighed into the automatic dispensing hopper of the mixer, 225 +/-1g water were placed into a dry mixer bowl. 450g anhydrous cement was weighed out, added to the mixer bowl and the automatic mixer started. The material was mixed at a slow speed for 30 seconds and the sand added over the following 30 seconds. The mix was allowed to rest for 30 seconds, the first 15 seconds of which any mortar stuck on the sides and bottom of the bowl was scraped back into the middle. Finally mixing was resumed at high speed for 1 minute. Cleaned and pre-greased Tonitechnik moulds were filled in two layers whilst on a vibrating table and vibrated for 120 seconds.

APPENDIX 2

Compressive Strength Determination of Concrete Cubes

The compressive strength of concrete cubes was tested at 28 days +/-4 hours. Having removed cubes from the curing tanks and wiped away excess grit and water the cubes were placed on their side and crushed using a Tonipact 3000 compression machine with rate of load set at 2.5 kNs^{-1}.

Compressive Strength Determination of Mortar Prisms

The compressive strength of mortar prisms was also tested at 28 days +/- 4 hours. On removal from curing the prism is broken in half then crushed compressively on their side faces using a machine compliant with clause 4.8 of BS EN 196 part 1. Load is applied at a rate of 2.4 +/- 200 kNs^{-s} and the failure load recorded.

ACKNOWLEDGEMENTS

The authors would like to express their appreciation for the financial support of the DTI and EPSRC through the Postgraduate Training Partnership (PTP) as well as the sponsorship and supervision of Rugby Cement.

REFERENCES

1. NEVILLE, AM. Properties of Concrete, 2nd edition, Pitman, 1973.

2. LEA, FM. Chemistry of Cement and Concrete, 4th Edition, Arnold, 1998.

3. MEHTA, PK. Concrete Structure, Properties and Materials, Prentice-Hall, 1986.

4. MITSUI, K, LI, Z, LANGE, DA, SHAH, SP. Relationship Between Microstructure and Mechanical Properties of the Paste-Aggregate Interface of Concrete. ACI Materials Journal, 1994, Vol. 91, pp. 30-39.

5. BENTUR, A. Microstructure, Interfacial Effects, and Micromechanics of Cementitious Composites. Proc. of Conference on Advances in Cementitious Materials, Gaithersburg MD, 1990, pp. 523-549.

6. MASO, JC. The Bond between Aggregates and Hydrated Cement Paste. Proc. of the 7th International Congress on Chemistry of Cement, Vol.1, Paris, 1980, pp. VII1/3-VII1/5.

STUDY OF STEEL –CONCRETE BOND IN ACI BEAMS UNDER MONOTONOUS AND CYCLIC LOADING

H Aoun
J Achouche
University of Guelma
Algeria
F Buyle-Bodin
University of Artois
France

ABSTRACT. This paper presents data from a study on A.C.I reinforced concrete beams under monotonous and cyclic loadings. The work carried out concerns the degradation of the steel-concrete bond. The experimental process adopted permitted to put stress on the phenomenon of cracking that intervenes during the damage of the steel-concrete bond under these two types of loadings. The real behaviour of a reinforced concrete structure depends largely on the steel-concrete bond. The study of this bond is often absent in the experimental and theoretical studies under the monotonous and cyclic loadings. By our contribution, we wish to bring some knowledge about the state of the stresses and strains of the steel and concrete, and their relationship with the bond strength and cracking of the structure elements.

Keywords: Concrete, Reinforcement, Bond, Anchorage, Beam, Monotonous loading, Cyclic loading.

Dr Hynda Aoun is Lecturer in Civil Engineering Institute, University of Guelma, Algeria. He specialises in the study of steel-concrete bond in structures.

Professor François Buyle-Bodin is Director of the Laboratory of Materials and Structures, University of Artois, Bethune, France. He specialises in the behaviour of reinforced fibrous concrete, shear strength of high performance concrete and steel-concrete bond.

M Jamel Achouche is Lecturer in Civil Engineering Institute, University of Guelma, Algeria. His main research is shear in reinforced concrete structures and shear wall structures.

INTRODUCTION

The choice of the A.C.I beam as a testing sample was a must. This is mainly due to the disadvantages brought by classical tests on bond, and to the non-conformity of these tests to explain a behaviour nearer to the actual one [1]. In order to approach the behaviour of a standard beam, we have modified the characteristics of the A.C.I beam by an increase of its length (from 2m to 3m), and particularly for the study under cyclic loading we have suppressed the reservations that exist in the lower part of the beam. The aim is to avoid the buckling of the reinforcement during the loading cycles on one hand, and to limit cracking on the other hand.

TEST PROGRAMME

Study of the Bond under Monotonous Loading

Characteristics of the testing specimen

The A.C.I beams were 3,050 m long and with a cross section of (205 x 457) mm². The main reinforcement is a H.Y steel (f_y=500 N/mm²) 16 mm diameter bar which extends beyond the limits of the beam, so as to be able to follow the slippage of the steel related to concrete during loading. The anchorage lengths at the ends of the beams are : 203 mm, 304 mm, 406 mm and 508 mm (Figure 1). A 'reservation' with dimensions 203 x 152 x 72 mm is provided in the lower part of the beam, in order to allow the measurement of the steel strain by the gauges put at this place, and to follow the evolution of the steel slip related to concrete during loading. The concrete used is of a strength about 30 MPa.

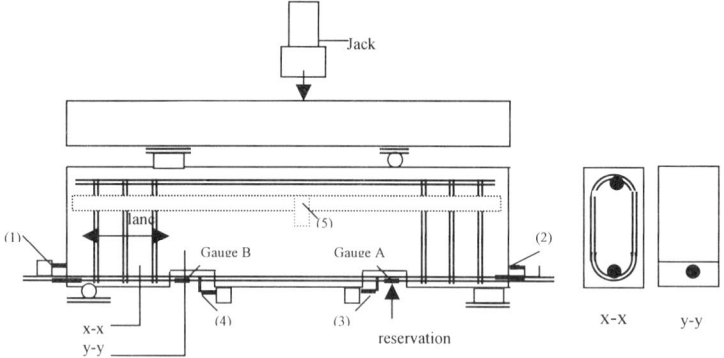

Figure 1 Geometrical characteristics, mode of loading and instrumentation

Test device and instrumentation

The A.C.I beams were loaded in four point bending (Figure 1). The load was applied by an INSTRON 250 kN jack. In order to follow the reinforcement slippage, we placed dial gauges (1) and (2) at both ends. The dial gauges (3) and (4) were placed at the level of the reservation to follow the slip between the central part of the beam and its ends.

The dial gauge (5) was to measure the deflection at mid-span of the beam. The gauges A and B measure the strain of the reinforcement during loading.

Results and analysis

The study of the curves (Figure 2) allows us to follow the evolution of deflection of the beams corresponding to the applied efforts. The beams behaviour presents a first phase which is perfectly elastic without a concrete damage. The second phase is non-linear and reveals a beginning of the concrete damage in the tensile zone and a loss of stiffness. Finally, a last phase of important plastic strains leading the beam to failure.

The geometry of the A.C.I beams is advantageous, because at the level of the main reinforcement for which we want to study the bond. We distinguish two parts : a loaded part and a free part (end of the beam). This allows us to evaluate the importance of the slippage of both sides of the steel. The curves (Figure 3 and Figure 4) show important slippage of the loaded side compared to the free side. The bar started slipping in a gradual manner on both sides. Then, we observed a sudden and important slippage without an increase of the load. A final phase which revealed a stabilisation of the slippage for certain reinforcements of the beams and a renewal of the slippage for others. The connection starts degrading first at the loaded side of the bar because it is nearest to the loading point, then the degradation progresses further until it reaches the free side. That is the result of friction phenomenon resulting from the failure of the steel-concrete bond. We have to notice that this complex phenomenon is amplified by the presence of the reinforcements crenellations.

The strains of the steels related to the applied forces (Figure 5) show a first zone representative of an elastic behaviour, followed by a second non-linear zone, which shows a loss of stiffness of the reinforcement under the applied forces. The curves of the bond stresses related to slippages (Figure 6) show the variation of the relative displacement between steel and the concrete in relation to the tensile forces in the reinforcement.

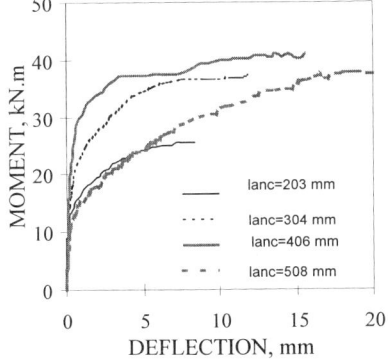

Figure 2 Moment-deflection relationship

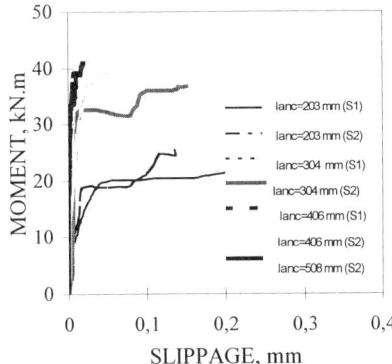

Figure 3 Slippage during the test (1, 2)

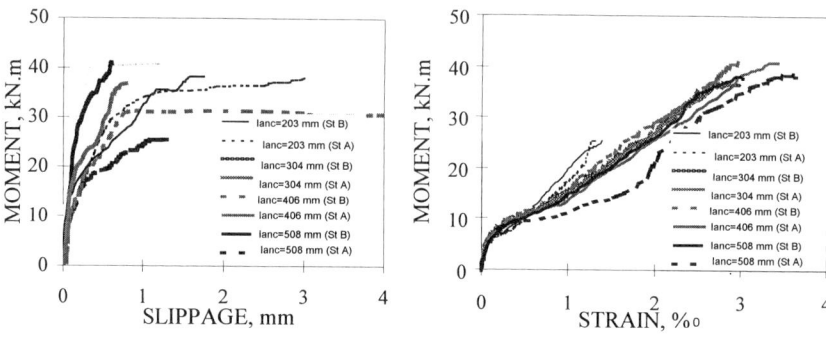

Figure 4 Slippage during loading   Figure 5 Strain during the test

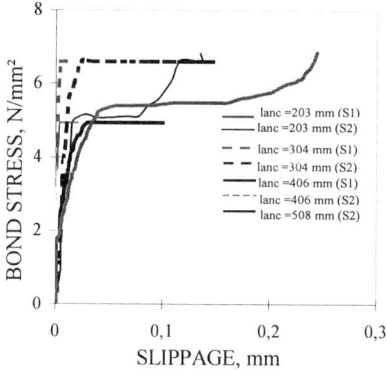

Figure 6 Bond stress related to the slippage

Study of the Bond under Cyclic Loading

Characteristics of beams, measurement and testing device, and loading program

The achieved A.C.I beams were 3,050 m long with a cross-section of (205 x 475) mm². For theses beams we have suppressed the 'reservations'. The main reinforcement is a H.Y 16 mm diameter bar. The upper reinforcement is symmetrical and of the same diameter to the lower one. The stirrups are 6 mm diameter HY bars placed at the beam ends. Their number is respectively 3, 4 and 5 for the beams B1, B2, and B3.

The A.C.I beams are loaded in an alternated bending. The beams were simply supported at both ends. The load is transmitted to the beam by an INSTRON 250 kN jack with a collar adjusted in the middle (Figure 7). The measurements of the reinforcements and concrete strains are assured by strain-gauges. The dial gauges record the alternated central deflexion of the beam and the slips of reinforcement at the extremities of this later.

The program of the cyclic alternated loading is defined from the results of the static tests. It is composed of phases with increasing amplitude and phases with decreasing amplitude starting from 45% up to 85% of the failure load (Figure 8).

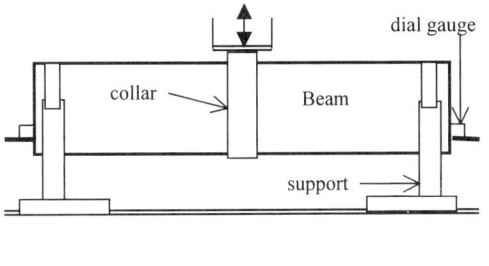

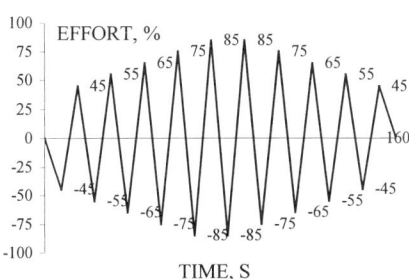

Figure 7 Testing device in alternated bending

Figure 8 Loading cycles

Results and Analysis

We have noted a certain stabilisation of the loops after the first weak cycles of hysteresis loops. The behaviour of three beams (Figure 9) is almost similar and is characterised by :

-The deflections of the beams during the first two cycles are small they become more significant from the third and fourth cycles.

-The curves of the beams B2 and B3 present a dissymetry in the behaviour of the strain. This dissymetry is influenced by the degradation of the beams during the inversion of the loading sign.

The evolution of the strain of the central zone highly loaded, indicates that the strains of the steel do not appear clearly until a certain number of cycles has been carried out. The important strains recorded are produced in a brutal manner and causing important slippage together with a loss of stiffness.

The curve (Figure 10) displays the irreversibility of the strains of the reinforcement, though theoretically it is in an elastic range stage. This phenomenon shows the steel-concrete interaction and the stresses acting at this level. This is available too for beams B1 et B2.

The results of these experiments reveal that the slippages of the reinforcement remain negligible during the quasi-dynamic loading [2] [4].

No evident failure of the steel-concrete bond took place, except for the curve in Figure 11 which shows clearly a slip of the reinforcement, it becomes progressively more significant with the loading cycles.

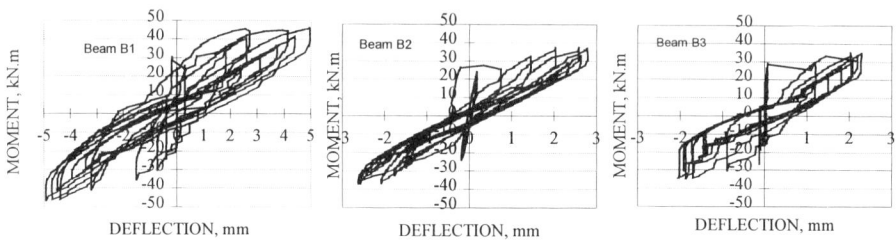

Figure 9 Moment-Deflection under cyclic loading of beams B1, B2 and B3

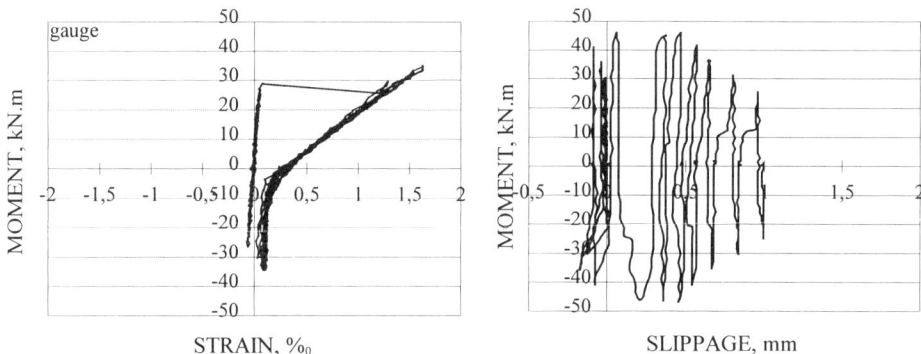

Figure 10 Strain of the reinforcement at mid-span of the beam B3

Figure 11 Slippage during loading of the beam B1

DISCUSSION

Behaviour of the Steel-Concrete Relationship

Under monotonous loading

At the first stage, the beam is not cracked as long as the tensile stresses in the concrete remain lower than the tensile strength (Figure 5). It concerns an elastic behaviour. Along the anchorage, the concrete deforms in an elastic manner while the steel slips. The second stage starts when the loading goes beyond the minimum value of the concrete tensile strength. The beam starts to crack. At this stage, the strain of the steel corresponds to an approximate value of 0,1‰. We note two phases :

A phase of formation of the cracks : characterised by the development and the appearance of new cracks. Therefore, the stiffness of the element decreased progressively. This phase stands until the obtainment of strains of the steel approaching 1‰.

The beginning of this phase emerges when the bar starts to wedge thanks to the crenellations. The bond stress is critical at this level since it highly depends upon this change of behaviour of the bar. In fact, after the first wedge of the steel, the slippage becomes less important. In return, the concrete forms a sheath around the reinforcement. Hence, the plastified concrete behaves like a rigid cylinder adherent to the reinforcement.

A phase of stabilised cracking : characterised by a quasi-elastic behaviour, there is no formation of new cracks. The strain of the steel varies in average between 1‰ and 2‰. The steel continues to slip and to strain. The concrete is totally plastified near the anchorages, and the bond stress is more or less stabilised.

Finally, the last stage is that where the failure of the beam occurs. Therefore, the curve of the behaviour is non-linear. The failure takes place when the reinforcement strain goes beyond the apparent limit of elasticity, that is a strain about 2.2 ‰.

Under Cyclic Loading

The results of the tests on the beam showed clearly that the steel-concrete relationship is disturbed when the reinforcement is pulled and pushed in an alternative manner. This bond is also influenced by the phenomenon of opening and closing of the cracks. It results in double cracking.

Alternated loadings do not necessarily accelerate the slippage of the reinforcement as far as the concrete is concerned. When the main reinforcement is stretched, the slippage are higher than those observed when it undergoes a compression when the loading sign is reversed. Damages are predominant when the main reinforcement of the beam is compressed. The phenomenon of lateral contraction of the steel reduces its bearing surface on the concrete in the case of tension, and increases this surface in the case of compression. As a result, the Poisson effect increases the deterioration of bond. We have noticed that this program of loading has limited the deterioration of the bond throughout the cycles. Consequently, we observed less important slippages but accompanied with an irreversibility.

During the first loading, two cracks appeared in the central tensile area of the beam followed by inclined ones The upper part undergoes a local compression. After inversing the sense of the loading, the same phenomena are observed on the lower part. The cracks in the upper part are closed again in this phase. Therefore progressively, in the cycles, the initial cracks become more and more open, and some side cracks are inter-crossed. The transmission of the effort takes place now in doubly cracked area which is less efficient for that [3].

CONCLUSIONS

The bond between steel and concrete does not result from adhesion of the steel to the concrete, but from friction. This friction is accompanied by the formation and development of cracks. Failure of the bond can be caused by failure in the concrete and simultaneously by slippage of the reinforcement.

Previous studies have clearly displayed the importance of the steel-concrete relationship in the global behaviour of the structure under cyclic loading.

This test is a first approach to the actual process of the behaviour of the steel-concrete junction under this type of loadings. The study has allowed the obtainment of curves of the global behaviour that characterise the decrease of the stiffness of the element, and local measurements that allow analysis of the phenomena of friction and cracking at the level of steel-concrete bond. The local measurements show that the steel-concrete junction is damaged by cyclic loading, and this leads to considerable local cracking.

Slippages remain small compared to deformations of the structure, indicating that this is not always the main source of the deformation of the structure. The results indicate that there was no clear failure of the steel-concrete bond under cyclic loading, and that the steel-concrete bond may resist loads nearer to those of the static failure applied to the concrete and to the reinforcement. Hence, it can be concluded that the failure of the bond intervenes after several loading cycles and that it is highly dependent upon the state of cracking. There also exists a close relationship between the steel-concrete bond behaviour and the opening and closing of the cracks.

REFERENCES

1. CHERVET, C. Propriétes d'adherence des armatures métalliques dans les ouvrages en béton. Nouvelles méthodes de caractérisation. Interprétation de l'adhérence en modèle élasto-plastique. Thesis of Doctorat, University of Nancy, 1978.

2. DEL TORO RIVERA, R. Comportement des nœuds d'ossature en béton armé sous sollicitations alternées. Thesis of Doctorat, ENPC, Paris, 1988.

3. GOTO,Y. Crack formed in concrete around deformed tension bars. ACI Journal, April 1971, pp 244-251.

4. ISMAIL, M A F, JIRSA, J O. Behaviour of anchored bars under low cycle over loads producing inelastic strains. ACI Journal, July 1972, pp 443-438.

INFLUENCE OF SILICA FUME ADDITION ON THE CONCRETE-REINFORCEMENT BOND

M Barbuta

University of Iasi

T Onet

University of Cluj-Napoca

Romania

ABSTRACT. The effectiveness of condensed silica fume (CSF) in improving the bond between concrete and steel has been investigated on two grade of concretes (with 325 kg/m^3 and 450 kg/m^3 cement content), with and without silica fume. The silica fume concrete included 0, 25, 40 and 60% silica as equal replacement of Portland cement for the Ist series and 20, 25 and 30% silica as equal replacement of cement for the IInd series. The pull-out method was used. The reinforcement were plain and ribbed. The tests showed that the use of CSF leads to an increase of the bond between the concrete and reinforcement. That is more significantly in the case of plain bars, when it can exceed 50%.

Keywords: Bond, Reinforcement, Condensed silica fume, Bond stress, Pull-out bar.

Dr Marinela Barbuta, is a Associate Professor in Reinforced Concrete, University of Iasi. Her main research interests include the use of additio in concrete, durability of concrete, corrosion of reinforcement.

Professor Traian Onet, is Chairman of Concrete and Civil Engineering Department, University of Cluj-Napoca. He specialises in prestressed and partial-prestressed concrete, durability of concrete and concrete structure, concrete repair, concrete with additions. Professor Onet has published widely and serves on many technical committee in the Romanian Concrete Association and in the European Committee for Standardisation.

INTRODUCTION

Condensed silica fume (CSF) is more and more used as admixture in high performance concrete, normal strength concrete and hydraulic concrete for improving different properties, such as: strength, durability or even fresh properties. But, for reinforced concrete members the use of silica fume results in the problem of additive effect upon the concrete-reinforcement bond, because the bond is the basis in using reinforced concrete as a structural material. The improved properties of concretes with CSF can offer the possibility to obtain an increase of the bond.

These initial investigations want to determine the effect of CSF on concrete-reinforcement bond. The tests were made according to our STAS 5511-89, i.e. the pull-out test. Experiments were realized on specimens (cylinders) of different grades of concretes, with and without CSF. The reinforcement used was type OB 37 (plain) and PC 52 (ribbed).

EXPERIMENTAL PROGRAMME

Materials

The concrete specimens were produced at the Concrete, Material and Technology Department of Civil Engineering Faculty, using the following materials:

Cement

The type of the cement was H II/A-S 42,5 SR 3011: 1996, the data on their physico-mechanical characteristics are given in Table 1.

Table 1 Cement characteristics

PHYSICO-MECHANICAL CHARACTERISTICS	CEMENT H II/A-S 42,5 SR 3011-1996
Initial setting time, min.	90
Expansion mm, min.	10
Hydration heat at 7 days, J/g, max.	280
Compressive strength N/mm^2:	
• initial at 2 days	10
• standard at 28 days	42.5 .... 62.5

Condensed silica fume

The silica fume (CSF) was obtained as a by-product resulting in the production of ferrosilicon alloys at Tulcea Plant, having the physical characteristics that are given in Table 2 [2].

Table 2 Chemical composition and physical properties of CSF

CHEMICAL ANALYSIS (%)					PHYSICAL TESTS		
SiO_2	Al_2O_3	Fe_2O_3	CaO	MgO	Density (kg/m³)	Sp. density (kg/m³)	Diameter μm
86-95	0,85-2,5	1,3-4	0,4-0,8	0,6-1,5	200-450	2000	0,5

Aggregates

Different batches of coarse and fine crushed aggregates from quarry of Poiana Teiului were used. The aggregates satisfy the conditions given by STAS 1667-76 and PE 713-90 and were separated in three sorts: 0 - 7 mm, 7 - 16 mm and 16 - 40 mm.

Air - entraining admixture

DISAN solution was used

Reinforcing steel

The following types of reinforcing steel rods were used:

- hot rolled plain steel (OB 37) of 8 mm diameter;

- hot rolled ribbed steel (PC 52) of 8 mm diameter.

Table 3 Test results for the bond between concretes with and without CSF and reinforcement

SERIES	CEMENT DOSAGE kg/m³		CSF kg		CSF %	
	Plain	Ribbed	Plain	Ribbed	Plain	Ribbed
I.1	325	325	-	-	0	0
I.2	243.75	243.75	81.25	81.25	25	25
I.3	195	195	130	130	40	40
I.4	130	130	195	195	60	60
II.1	450	450	-	-	0	0
II.2	360	360	90	90	20	20
II.3	337.5	337.5	112.5	112.5	25	25
II.4	315	315	135	135	30	30

The control concrete mixes had the following proportions in 1 m³:

- (I) cement, 325 kg; fine aggregate 846 kg; coarse agregate 1077 kg; water, 199 l; and superplasticizer, 3.25 l; different silica fume contents of 25, 40 and 60% by mass of cement, replaced an equal weight of cement in the mix.

- (II) cement, 450 kg; fine aggregate 818 kg; coarse aggregate 998 kg; water, 128 l; and superplasticizer, 4.5 l; different silica fume contents of 20, 25, and 30% by mass of cement, replaced an equal weight of cement in the mix.

Test Specimens

For the determination of the bond by the pull-out tests, three specimens were prepared for each series from the two kinds of concrete. For each series there were samples with plain steel and with ribbed steel. The samples were cylinders of 320 mm height and 159 mm diameter.
The test specimens designed for pull-out testing were taken out of their forms after 48 hours, and up to their testing they were kept at a temperature of 20 ± 3°C in open air with a relative humidity of 65% ± 5%.

RESULTS AND DISCUSSION

The data obtained from the experimental tests are given in Figure 1, 2, 3.

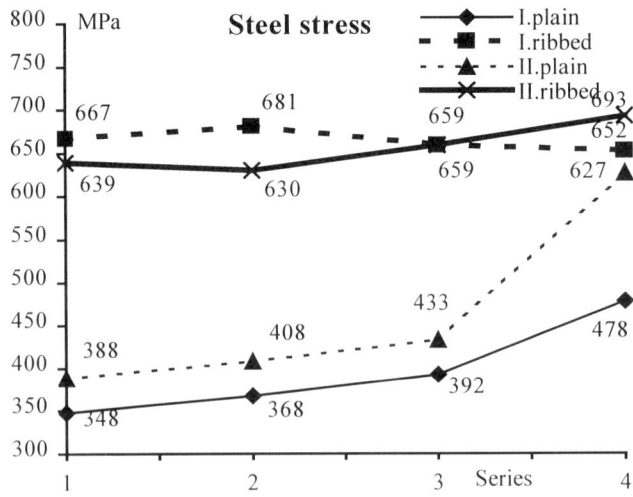

Figure 1 Stress in steel for series 1-4

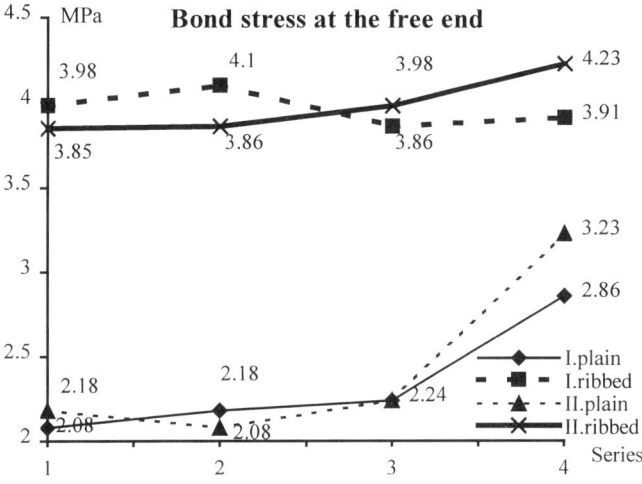

Figure 2 Bond stress at the free end for series 1-4

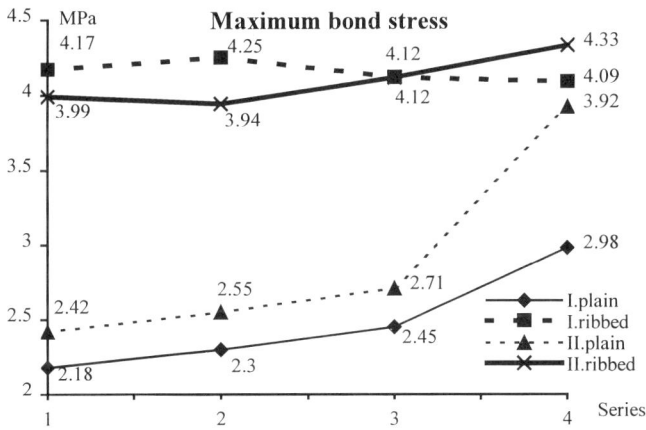

Figure 3 Maximum bond stresses for series 1-4

Note: $\tau^{0.01}$, τ^{max} are the stresses of the bond with the free end of the reinforcement displaced by 0.01 mm and for the maximum force, respectively.

Analysing the experimental results we can observe that the CSF has increased the bond strength. That is seen to be more strongly in the case of plain reinforcement (series I.2 plain, I.3 plain, I.4 ribbed, II.2 plain, II.3 plain, II.4 plain).

In this case because the concrete mixture with CSF is less liable to segregation and water separation [1], the water layer under the reinforcement is reduced, that increases the contact zone surface between the concrete and reinforcement. With the increase of the CSF content we observe an increase of bond strength (for the mixture series I with 60% CSF the bond strength increase is maximum and represents 37.5% and for series II the maximum increase of bond strength is obtained for mixture with 60% CSF and represents 48.1%).

In the case of ribbed reinforcing bars the experiments carried out have determined a maximum increase of bond strength in the case of mixture with 25% CSF for the series I and in the case of 30% CSF for the series II. In the case of Ist series only for the mixture with 25% CSF the bond strength increased compared to the control specimen (without CSF), but the other values were smaller, but very close to the control value. In all cases of series II the results show that the CSF additive has increased the bond strength. It must noted here that a considerable deviation exists in the individual results of each series, as well as a substantial difference in CSF contribution on the bond strength.

About the maximum steel stress it can be observed that in the case of concretes with CSF it has bigger value than the control value, but also here it must be noted that the results were influenced by some technological factors and because of that in some cases the reinforcement had broken before the bond was destroyed.

CONCLUSIONS

From the test results presented here, the following conclusions may be made for the influence of CSF on the bond between concrete and steel.

The use of CSF as cement replacement in concrete leads to an increase in bond strength. This increase is more significant in the case of plain reinforcement, when the increase can go to 50% and more. If the cement dosage is small, higher CSF content higher increase of bond strength. In the case of concretes with high cement dosage, is not the same rule and the optimum proportion for maximum bond strength is obtained for smaller CSF content.

The CSF improve also the maximum bond stress, that means the influence of CSF is more significant and all aspect of bond such as friction, adhesion and mechanical interlocking must be studied in wider investigations.

REFERENCES

1. NEVILLE, A.M. Properties of Concrete. Fourth Edition, Longman Group Limited, Essex, England 1995, pp 844.

2. BARBUTA, M. and PATRAS, M. Properties of hydraulic concrete incorporating large amounts of silica fume. Technical University Magazine Iasi 1997, volume XLV, section VI, pp. 29-35.

3. ***, STAS SR 3011 – 1996, Limited hydratation warmth cements and resistant to water damage with sulphates content.

4. ***, STAS 5511 – 1989, Pull-out test.

INTERFACE STRENGTH IN POROUS POLYMER CONCRETE

A Beeldens
D van Gemert
K U Leven University
Belgium

Y Ohama
K Demura
Nihon University
Japan

L Czarnecki
Warsaw Technical University
Poland

ABSTRACT. Porous concrete is a polymer modified cement concrete with high accessible porosity, developed as a top layer for highways. The strength of porous concrete is provided by the polymer cement co-matrix and therefore dependent on the strength of the matrix and on the adhesion strength between coarse aggregate and mortar. This paper presents the results of a test programme to investigate the influence of the polymer-cement ratio and curing conditions on the adhesion strength and other properties of the polymer-cement mortar. The adhesion strength was determined on real scale samples by means of a direct tensile test. In addition, a SEM-study was carried out to reveal the structure of the polymer modified mortar and the presence of polymer film.

Keywords: Porous concrete, Adherence, Polymer-cement ratio, Curing conditions

Ir Anne Beeldens is a research assistant at the Department of Civil Engineering of K.U.Leuven. She is doing a PhD concerning the structure and the behavior of porous concrete: the influence of polymer modification on the mechanical properties and durability of concrete, related to the microstructure of the polymer-cement mortar.

Dr Dionys Van Gemert is professor of building materials science and renovation of constructions at the Department of Civil Engineering of K.U.Leuven. His research concerns repair and strengthening of constructions, concrete polymer composites.

Dr Y Ohama, honory doctor of K.U.Leuven, is professor of Architecture at the College of Engineering, Nihon University, Koriyama, Japan. He has been involved in the research and development of concrete-polymer composites for more than thirty-five years.

Dr K Demura is associate professor of Architecture at the College of Engineering, Nihon University, Koriyama, Japan. His main interest has been in research and development of concrete-polymer composites.

Dr L Czarnecki is professor of polymer building materials at Warsaw Technical University. His research concerns application and optimization of concrete-polymer composites.

INTRODUCTION

Porous concrete is a polymer modified cement concrete with high porosity. Its application as top layer on highway roads requires high strength (up to 18 MPa compressive strength and 4 MPa flexural strength) at large water accessible porosity (25 %). The structure of porous concrete is composed of three phases: aggregates, which are discontinuously dispersed in the material, a polymer cement co-matrix mortar and large pores, which are water accessible and form a second matrix through the material [3, 4]. The mortar contains the polymer cement paste and the small amount of sand. The strength of porous concrete is provided by the polymer-cement co-matrix. This co-matrix encapsulates the aggregates and binds them together by means of mortar bridges. The layer thickness around the aggregates varies from 0.5 mm to several mm.

Tests are carried out on the polymer-cement co-matrix, to investigate the influence of polymer-cement ratio and curing conditions on the strength of porous concrete, and on aggregate-mortar samples to reveal the strength of the bulk matrix and the aggregate-mortar adhesion strength.

Because the mixture composition of porous concrete does contain a very small amount of fine aggregate, large volumetric shrinkage of the matrix occurs. Therefore, traditional measuring techniques developed for repair mortars, can not be used. A real scale composite sample is prepared and a direct tensile test is carried out to reveal the adhesion strength and the tensile strength of the matrix. Additionally, a SEM-study is made to relate the real structure, the polymer film formation and the cement hydration to the properties of the material.

MATERIAL PROPERTIES

Properties of the Bulk Matrix

Five different cement mortars were made with varying polymer-cement ratio, to investigate the influence of the polymer addition. According to the mix composition of porous concrete, a small amount of fine sand (0-1 mm) was used: sand-cement ratio equal to 0.16 (by weight). The polymer-cement ratio was taken equal to 0 %, 5 %, 10 %, 15 % or 20 % (weight of solid phase of polymer emulsion by weight of cement). Polystyrene-acrylic ester latex with solid phase equal to 50 % was used. The water-cement ratio was equal to 0.3, taken into account the water from the polymer emulsion.

Three different curing conditions were imposed: standard curing (2 days at 95 % R.H. and 20 °C, 5 days in water at 20 °C and 21 days at 60 % R.H. and 20 °C), water curing (2 days at 95 % R.H. and 20 °C and 26 days in water at 20 °C) and dry curing (2 days at 95 % R.H. and 20 °C and 26 days at 60 % R.H. and 20 °C). After 28 days all remaining samples were stored at laboratory conditions (60 % R.H. and 20 °C).

The flexural strength and compressive strength of the bulk mortar matrix were determined on beams (40 x 40 x 160 mm), according to standard NBN B12-208 (Belgium Standard). The direct tensile strength was determined in a similar way as the adhesion strength between aggregate and polymer cement co-matrix.

Due to the small amount of sand, large shrinkage occurred in the samples. The shrinkage was measured according to AFNOR P18-361 (French Standard). The total shrinkage increased with increasing polymer-cement ratio. Table 1 gives the total shrinkage after 28 days for the different curing types at different polymer-cement ratio.

Table 1 Total shrinkage after 28 days (µStrain)

CURING	POLYMER/CEMENT (P/C) RATIO, %				
	0	5	10	15	20
Standard	-18.04	-16.20	-26.04	-33.01	-35.16
Water	4.10	4.30	9.22	14.56	19.58
Drying	-22.79	-21.42	-27.27	-41.00	-79.75

The rate of shrinkage was the highest just after the samples were placed in a dry environment. For dry cured samples this was after 2 days, for standard cured samples after 7 days and for water cured samples after 28 days. In moist or water conditions, the samples absorbed water molecules and increased in volume. The water absorption was higher with increasing polymer-cement ratio.

The flexural strength results for the standard cured samples and for the dry cured samples are shown in Figure 1.

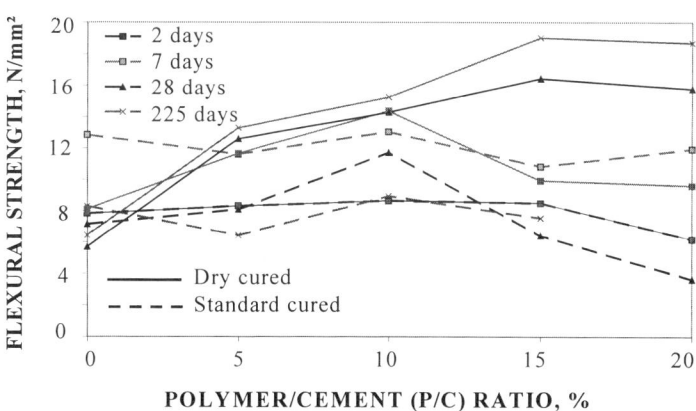

Figure 1 Flexural strength for standard cured samples

Remarkable was the fact that the bending strength of the standard cured samples after 7 days was much higher than the bending strength of the samples after 28 days and after 225 days. This could be attributed to the large shrinkage, which occurred at the beginning of the drying period after 7 days of curing. During conditioning in water, cement hydration took place and a cement matrix was formed. Subsequent drying of the samples caused a large decrease in volume and induced micro-cracks in the material. The cement structure was partly destroyed and the flexural strength of the mortar diminished. Although shrinkage increased with increasing polymer cement ratio, the diminishing of the flexural strength was the smallest for mortar with 10 % polymer-cement ratio. The compressive strength was less influenced by the shrinkage.

The flexural strength for the dry cured samples at different curing ages is also presented in Figure 1. In this case, the drying period started after 2 days. Apparently the flexural strength was not influenced by the large shrinkage. With the dry curing conditions, cement hydration simultaneously took place with polymerisation and with drying shrinkage. Therefore the stresses were limited and the polymer film could bridge the micro-cracks. In general, the bending strength increased with increasing polymer-cement ratio with a maximum around 15 % polymer-cement content.

The addition of polymer emulsion seemed to act as retarder for the hydration of cement and polymer film formation, especially at higher polymer-cement ratio (15 % to 20 %). This is visible in the results of the dry cured samples, where the flexural strength of the samples with 15 % and 20 % polymer-cement ratio is much lower than the flexural strength of the samples with lower polymer-cement ratio. After 28 days, the flexural strength of the samples with 15 % to 20 % is higher than that of the samples with lower polymer-cement ratio.

Properties of the Aggregate-Mortar Composite Sample

Due to the small amount of fine sand in the mortar, large shrinkage occurred during the hydration process, as shown in the previous part. This large deformation made it impossible to measure the adhesion strength on large samples. Therefore real scale samples were made.

The composite samples were formed by dipping a cube 10 x 10 x 10 mm, cut of bulk aggregate, into fresh mortar, so a thin layer was deposited on the cube. The thickness of the layer decreased with increasing polymer-cement ratio, and varied between 0.4mm (15 % p/c) and 2.6 mm (0 % p/c). After curing, the upper and bottom surface were slightly scrubbed with abrasive paper P150 and two traction bolds were glued at opposite faces with a two component epoxy glue. These bolds were positioned in a tensile machine by means of two hinge joints. The tensile test was carried out at a speed of 1 mm/minute. The adhesion strength was calculated and the adhesion surface characterised.

RESULTS

Adhesion Strength of Aggregate-Mortar Composite Samples

The adhesion strength of the aggregate-mortar composite sample was tested as described before. Due to technical problems (decohesion of the glue, failure between mortar and tension bar, torsion during positioning of the sample in the press) some samples failed without relevant results.

Figure 2 gives the adhesion strength after 58 days curing, averaged over at least 5 samples of the same mixture and the same curing conditions. Figure 2 also presents the direct tensile strength of the pure mortar, measured under the same methods.

Each time, the type of fracture was characterised. Two different types were distinguished: rupture in the mortar layer and in the interface between mortar and aggregate. For the samples with no polymer modification rupture took place at the transition layer mortar-aggregate. A similar tendence was seen for all the water cured samples.

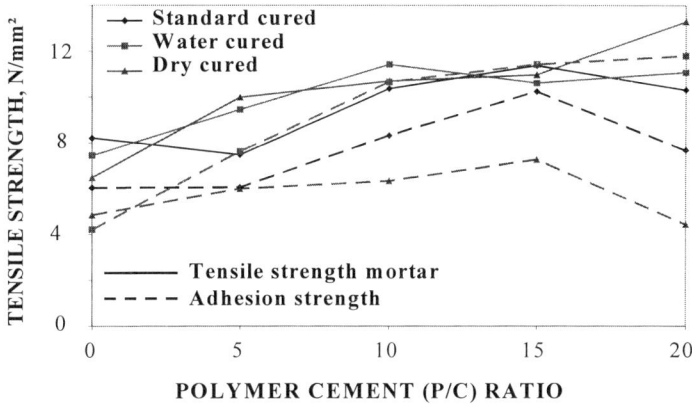

Figure 2 Adhesion strength between aggregate and mortar after 58 days curing

SEM-Study of the Fracture Surface

The fracture surface of the mortar and of the composite sample aggregate-mortar was investigated by means of a scanning electron microscope (SEM). Figure 3 reveals the structure of the mortar with p/c-ratio equal to 10 %.

In this figure the simultaneous growth of cement hydrates and polymer film is clearly visible, although the polymer particles have not completely coalesced together to form a continuous polymer film.

Figure 4 shows a $Ca(OH)_2$ crystal at the surface of mortar with 0 % polymer-cement ratio. Figure 5 represents a similar $Ca(OH)_2$ crystal at the surface of mortar with 15 % polymer-cement ratio.

In the first case, small CSH-needles are visible and a large $Ca(OH)_2$-crystal is visible. In the other case, the presence of polymer particles that have partly coalesced in a polymer film is clearly visible. No small needles are present and the crystal structure is not developed completely.

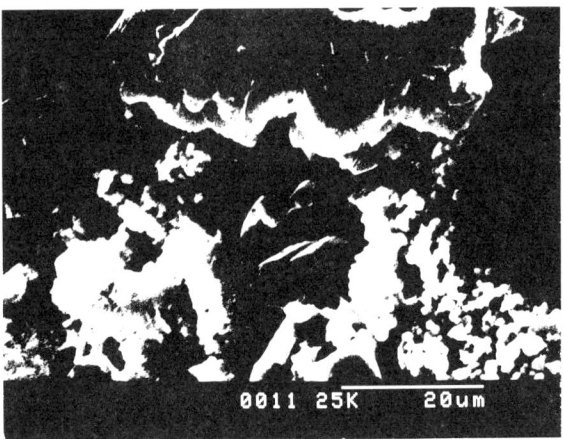

Figure 3 Fracture surface of the mortar with 10% p/c ratio

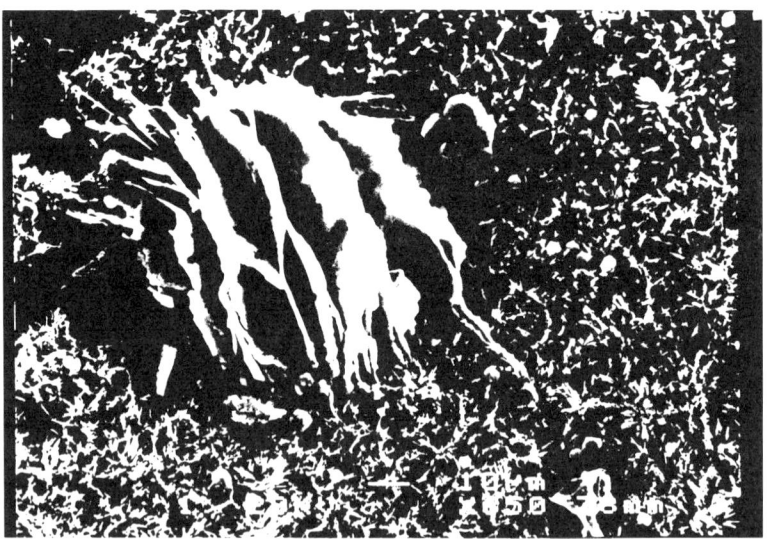

Figure 4 Surface of the mortar with 0% p/c ratio

Polymer modification is believed to increase the tensile strength and the flexural strength of the bulk mortar and to ameliorate the bond between mortar and aggregate. This property is attributed to the polymerisation and coagulation of the polymer particles into a continuous film, which interpenetrate with the structure of the hydrated and unhydrated cement particles. A chemical reaction between the silicate, the calcium hydroxide and the ester group of the polymer also increases the bond between matrix and aggregate [1].

Figure 5 Surface of the mortar with 15% p/c ratio

Although some SEM-investigation is made from the fracture surface of the composite samples mortar-aggregate, no clear distinction between the different types of failure nor between the different types of curing has been made.

DISCUSSION

Before commenting on the results, some remarks will be made concerning the test procedure. The advantage of this method was that small-scale samples were used, simulating reality. Hence, large shrinkage strains could be avoided and preliminary release of the mortar was avoided. On the other hand, the size of the sample gave difficulties in preparing and handling the samples. Compaction of the mortar layer on the aggregate was not possible and also the thickness of the mortar layer could not be controlled and diverged with varying polymer-cement ratio. This might have influenced the adhesion strength and the type of failure.

The execution of a direct tensile test is complex since a small angle in the sample induces a bending moment and will apparently lower the adhesion or tensile strength. This problem was for the greater part corrected by the installation of the hinge joints in the press and by working with aggregate with parallel sawn surfaces. Due to this, only sawn surfaces were tested at this stage.

Another problem, which occurred during the test, was the fixation of the sample to the bolts, more particularly the adhesion between the mortar layer and the glue. To provide an adequate adherence the mortar surface had to be dry and rough. Hence, it was necessary to dry the water-cured samples prior to testing and to scrub the mortar surface with abrasive paper. This could induce micro-cracks in the material.

The formation of the polymer-cement co-matrix is influenced by the curing conditions. Cement hydration is favoured by wet curing conditions, polymer film formation on the other hand takes place primarily in dry curing conditions. Therefore curing conditions that consist out of a wet curing period, followed by a dry curing period provided optimum conditions. The result indicated that the strength development was higher when the wet curing period was longer, as is indicated in figure 3. 28 days water curing, followed by a dry curing period (water curing) gave better results than 7 days water curing followed by a dry curing period (standard curing). In the first case, cement hydration could take place extensively, followed by polymer film formation.

From Figures 2 it follows that bond strength and tensile strength indeed increase with increasing polymer-cement ratio. The type of failure also indicates a better bond between the mortar and the aggregate when polymer emulsion is added to the mixture.

Figures 3, 4 and 5 show the development of cement hydrates and polymer film interpenetrating each other. Polymer modification reduces the formation of large crystals which is also favourable for the strength development since large crystals have more orientated surfaces which serve as preferred cleavage site. [2]

CONCLUSIONS

The development of a direct tensile test on real size composite specimens made it possible to determine the bond strength between mortar and aggregate for porous concrete.

Polymer modification improves the adhesion strength and the tensile strength of the mortar due to the collaboration of cement hydrates and polymer film. The influence of curing conditions was clearly visible. Dry curing conditions favoured polymer film formation, wet curing conditions promoted cement hydration. Optimum curing conditions were found to be 28 days water curing followed by 30 days at dry conditions.

REFERENCES

1. OHAMA, Y. Handbook of polymer-modified concrete and mortars, properties and process technology. Noyes Publications , 1995, p. 236.

2. MEHTA, PK AND MONTEIRO, PJM. Concrete - Structure, properties and materials. Prentice Hall, 1993, p. 537.

3. BEELDENS A, VAN GEMERT, D, CAESTECKER, C (LIN), AND VAN MESSEM, M (LIN). Structure and Performance of Porous Concrete. Proceedings of 10th ICCC, June 1997, Gothenburg, p. 4iV007.

4. BEELDENS, A, VAN GEMERT, D, OHAMA, Y AND DEMURA, K. Strength and durability of porous polymer cement concrete. Proceedings of the IXth ICPIC, pp 14-18 September 1998, Bologna.

MECHANICS OF FAILURE OF RIGID APPLIED FINISHES TO CONCRETE SURFACES

P Dux

University of Queensland

P Mullins

Mullins and Associates

Australia

ABSTRACT. This paper addresses the failure of rigid finishes such as tiles, cut stone and fired clay pavers adhered to concrete surfaces via mortar or equivalent adhesive. The paper describes the mechanics of failure, showing that both the short term and time-dependent behaviour of the concrete, the mortar and the rigid finish material are of significance. The paper illustrates stress regimes established within materials and at interfaces under volumetric changes from thermal and moisture variation. From these regimes, good and bad practice can be identified. For example, thin mortar beds and incomplete founding of the finish units in the mortar bed lead to higher stresses and increased risk of failure. It is shown that boundary zones of high shear stress which can develop, are localised and are largely independent of the size of the tiled or paved area; that is, as volumetric change occurs, there is no gradual accumulation of shear stress with distance from the centre of the work. Amongst other findings, it is shown that movement joints can create boundary conditions conducive to failure, thus defeating the purpose of their installation.

Keywords: Rigid finishes, Claddings, Tiles, Pavers, Failure, Debonding, Facades, Floors, Adhesive failure.

Dr Peter Dux is a Reader in Civil Engineering at the University of Queensland, Australia. His areas of research and consulting include concrete technology, concrete structures and stability of steel structures. He was awarded the 1995 T Y Lin prize by the ASCE, for research into prestressed concrete. He is Vice-President of the Queensland branch of the Concrete Institute of Australia.

Dr Peter Mullins conducts a consulting practice and is a part-time university lecturer. He is engaged mainly in building design and assessment of concrete structures for durability and repair. His principal research area is strength and serviceability of masonry structures.

INTRODUCTION

Rigid adhered finishes are often applied to concrete surfaces. Examples are tiles and other rigid cladding on walls, tiles and pavers on floors, and coping pavers and tiles on concrete pools. A strong element in the choice of surfacing product is appearance. Unfortunately, the engineering properties of the surfacing may not be suited to the application. Occasionally, failure occurs and part or all of the finish comes away from the concrete. This may involve loosening over large areas. It may feature the development of drumminess, peaking or heaving away from the boundaries of the area. There may be obvious reasons such as poor workmanship but sometimes, failure occurs despite the adoption of what is considered to be good practice.

Through numerical studies of basic models, this paper describes the interaction between substrate and applied finish and from this, identifies the mechanisms of failure. While, for convenience, the paper focuses on the behaviour of fired clay paving and tiling, the underlying principles are broadly applicable. The findings apply in situations where the finish adheres to the substrate via a mortar bed or some stiff adhesive as opposed to one that offers little resistance to relative movement between the finish and the substrate.

BASIC PROPERTIES OF THE MATERIALS

Fired Clay Materials

Ceramic tiles and fired clay paving tend to naturally grow with time by taking in moisture. Paver growth is classified as slight (< 0.6 mm/m), medium (< 1.2 mm/m) and high (> 1.2 mm /m) [1]. Most manufacturers can quote growth values for their products. Of greater importance are the stresses generated by the restraint of growth, and this information is generally not available.

Properties such as elastic modulus vary enormously. Tests by the authors on fired clay pavers gave elastic moduli ranging from 5 GPa to 20 GPa, the stiffer moduli being for pavers with a known poor performance history of delamination from concrete substrates. Hendry [2] lists a similar range of short-term moduli for brick masonry. Creep in clay based products is known to be less than that in concrete and mortar [2] but most tests relate to brickwork as opposed to the clay products themselves. A creep factor (creep strain/elastic strain) of around 0.5 is probably reasonable.

Concrete and Mortar

Elastic modulus and creep and shrinkage characteristics concrete are described in codes of practice. For building grades of concrete at the lower end of the strength scale (25 MPa – 32 MPa characteristic cylinder strength), average values for elastic modulus and creep factor are around 27 GPa and 2.5 [eg. 3]. The tendency for concrete to shrink depends on environment and physical proportions of the cast object [3].

The basic properties of mortar vary widely, partly due to site mixing. Mortars used as adhesive layers tend to be rich in cementitious material. Reference 4 quotes moduli of ranging from 1.7 GPa to 12.6 GPa. The creep factor for mortar could be expected to exceed that of concrete because of the absence of coarse aggregate (say, 3.0 – 4.0).

The tensile strength and shear strength at interfaces vary widely [2,5] depending on numerous parameters. Conservative estimates are 200 kPa and 250 kPa respectively although the range of test results is broad [2, 5, 7].

All of the materials respond to temperature change, the coefficient of thermal expansion being reasonably uniform at around $10 \times 10^{-6}/C^\circ$. Temperature changes may be seasonal or may occur very rapidly. Natural changes in size due to gain or loss of moisture occur gradually, hence stress effects from this develop slowly. A rapid change gives rise to stresses higher than those generated by an equal but slow change, as the latter are alleviated by the effects of creep. It follows that a small dimensional change, rapidly generated, can have more significant consequences than a much greater but gradual change.

FUNDAMENTAL BEHAVIOUR

Deformations and Stresses

This section highlights qualitative aspects of behaviour which apply in a wide range of situations. The simple, plane strain model [6] in Figure 1 is of a strip of surfacing units adhering via a mortar bed to a relatively massive concrete substrate. Real examples include tiles set in a mortar bed, or equivalent, on a concrete floor and coping pavers on the ring beam of a swimming pool. If the substrate is massive in relation to the surface finish the tendency of the surfacing to expand causes little sympathetic movement in the substrate. Similarly, while the surfacing might influence the tendency of the substrate to shrink by modifying the hygrothermal regime, it offers negligible mechanical restraint. The model does not identify grouted joints nor does it allow for flexibility in the substrate. These are considered later.

In the model, the somewhat arbitrary choice is for the surfacing to be fired clay pavers, 30 mm thick. The elastic modulus is 15 GPa. The mortar layer thickness, t_m, is either 10 mm or 20 mm as indicated in Figure 2. The paver-mortar modular ratio, n, varies from 2:1 to 20:1. The load case is a differential average temperature increase of 10 C° within the paving with respect to the mortar and substrate, chosen to cause expansion in the paving. The load case is not entirely arbitrary – the authors are aware of debonding of tiles from concrete substrates in spa pools, coincident with rapid heating of water over ranges of 10 – 15°C.

Figures 1(a) and 1(b) are elevations of the computer model. A coarse model has been used for clarity. A very fine model is required to provide sufficient data for plotting interpolation to reproduce a drop to zero shear stress at the free face. However, a few centimetres away from the free face, coarse and fine models predict essentially the same behaviour.

The left-hand end of the model represents a physical end at a movement joint across which the substrate is continuous (a typical construction detail); the right hand end is the centre-line. Figure 1(b) shows before and after shapes, with deformations magnified for clarity. The most notable feature in Figure 1(b) is that little deformation happens except at the physical end (left hand end). The paving expands vertically as expected, but horizontal movement is limited largely to the end region, most of the paving being restrained horizontally. The compressive forces required to do this are generated near the physical ends and involve only short lengths of paving interacting, via the mortar, with the substrate.

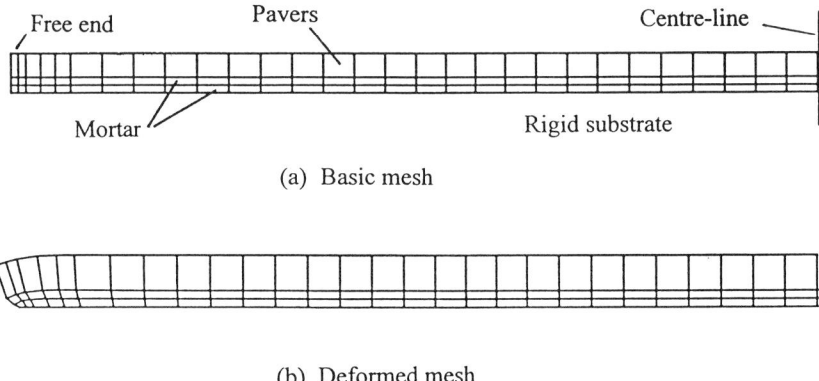

(a) Basic mesh

(b) Deformed mesh

Figure 1 Computer model of paving on massive substrate

Figures 2(a) and 2(b) show distributions of stress and identify some parameters of influence. Figure 2(a) shows shear stress distributions at the mortar-paver interface for different thicknesses of mortar and different modular ratios. Even with the 20:1 modular ratio, the region of high shear remains within a few hundred millimetres of the physical end of the paving. Distributions of horizontal stress within the paving are plotted in Figure 2(b). Compressive stresses are seen to develop over short distances near the physical ends and to remain reasonably constant over the remainder of the run. Not shown are vertical stresses across the mortar-paver interface which feature tension at the free face, adjacent zones of compression and tension within 200 mm of the free end, and negligible stress over the remainder of the interface.

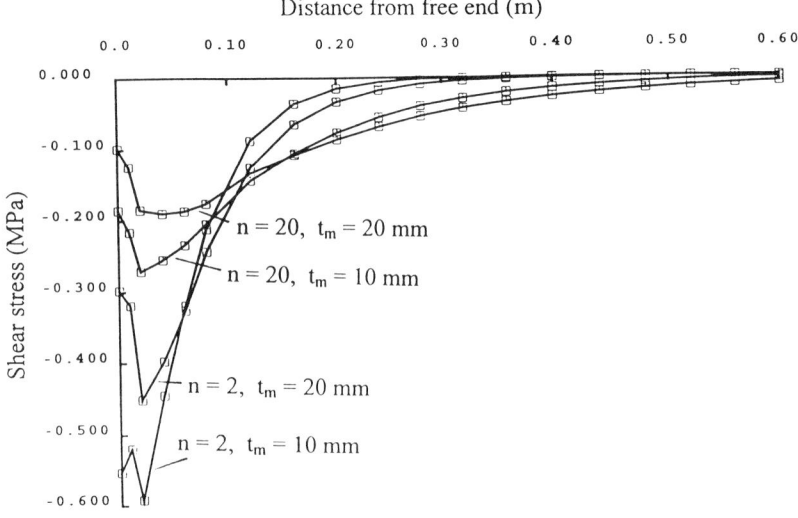

Figure 2(a) Shear stress at paver-mortar interface

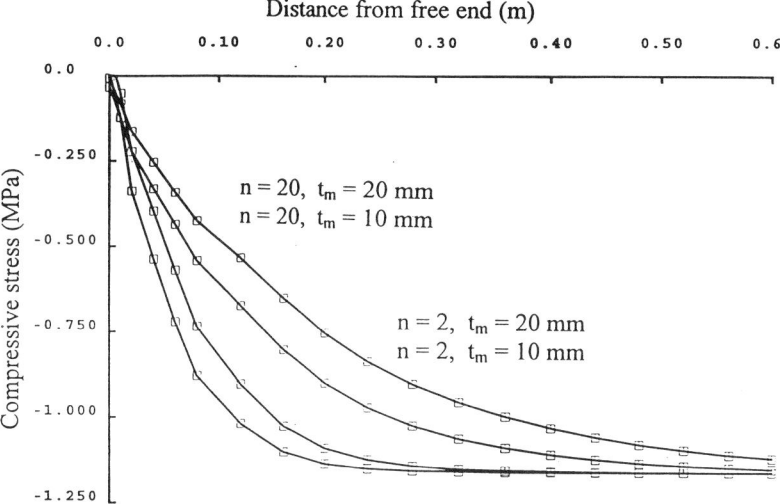

Figure 2(b) Horizontal compressive stress at mid-height of pavers

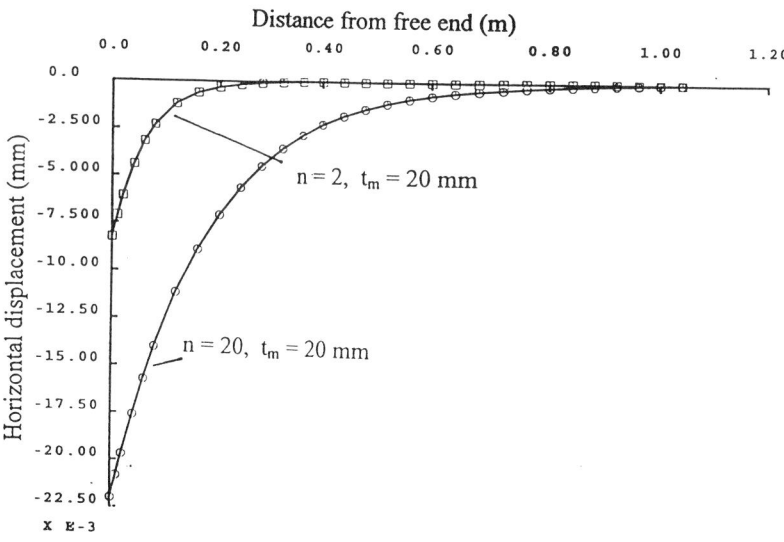

Figure 2 (c) Horizontal displacement of pavers at mid height

Figure 2(c) shows the average longitudinal displacement of the pavers, the negative sign indicating movement towards the free end. At the free end, the movements are 8% and 20% of the free expansion of the paver, accumulated from the centre-line, for modular ratios 2 and 20 respectively. At a distance of 400 mm from the free face, the actual longitudinal displacements reduce to about zero for modular ratio 2 and to 4% of the accumulated free displacement at that location for modular ratio 20.

The figure indicates that most movement occurs near free ends and that, beyond very short runs, the amount of relative movement is largely independent of the length of the run. Changes such as surfacing unit properties, grouted joints and substrate flexibility modify the above results but key qualitative features are retained. The simple model leads to the following observations:

- Similar, localised end conditions develop in short runs and long runs. The notion that shear strain and stress in the mortar accumulate uniformly with distance from the centre of a run or centre of an area is erroneous. Accumulation of strain can occur if the mortar is replaced by very flexible adhesive, and provided grouted joints between units are stiff.
- Delamination of the surfacing can initiate in the end zone due to excessive shear stress combined with local biaxial tension.
- Because the materials are brittle, delamination can propagate over virtually the full length of the run. Progressive failure, in which the highly stressed zone moves from the periphery towards the centre, will cease only when the remaining zone of bonded surfacing is small. It is therefore a common occurrence for large areas of tiling and paving to delaminate rather than just small areas.
- The compressive stresses in the surfacing combined with undulations in level can give rise to tensile stresses across the mortar-surfacing and mortar-substrate interfaces. This can lead to vertical separation (peaking and heaving failure) in the zone away from boundaries, a condition often detected by drumminess of the surfacing.
- A reduction in mortar thickness leads to higher interface stresses as the mortar becomes stiffer in shear. Good practice calls for thicker rather than thinner adhesive layers.
- A reduction in mortar stiffness results in reduced interface stresses. However as the shearing deformation of the adhesive in the end zone is increased, increased demand is placed on the adhesive bond to remain effective under higher shear strain.
- Incomplete bedding of the surfacing units gives rise to increased shear strains and stresses in the adhesive. The effect is equivalent to that of increasing the modular ratio by increasing the modulus of the surfacing units. Good practice calls for complete bedding of tiles and pavers in the adhesive layer.

The question arises as to the effect of substrate shrinkage. For the simple model, the overall result is much the same – shear and tension arise in the mortar at end region and most of the paving is placed in compression as it is forced to comply, in the main, with the preferred deformed state of the massive substrate.

Movement Joints

Conventional practice is to place movement joints at regular intervals in tile and paving work. The joints extend in depth to the top of the substrate. The concept is that such joints prevent the accumulation of shear strain in the adhesive layer. However, unless the adhesive is very flexible, accumulation of strain and relative displacement between the surfacing and the substrate do not occur.

Rather, as discussed above, end or boundary conditions are established which are largely independent of the extent of the area.

Obviously, joints through the mortar and surfacing should be placed over joints in the substrate. Other movement joints might be used isolate features such corners in coping work which are known to be prone to failure. However, the simple models suggest that regularly spaced movement joints are more likely to do harm than good if the adhesive is not very flexible, as high stresses are established in the work on either side of the joint. As the total length of movement joint increases so does the total area of highly stressed end zone. The authors have observed paving failure to initiate at movement joints.

BEHAVIOUR OF MORE COMPLEX MODELS

Figure 3 (a) shows a more complex plane-strain model with a finite substrate and grouted joints between pavers. The construction features 30 mm fired clay pavers of length 195 mm and with 10 mm joints (2 layers of elements in Figure 3), on a 20 mm mortar bed (2 layers of elements) over a 150 mm concrete substrate (3 layers of elements). The model represents one half of a slice a little over 3 m in length. The left-hand end is a free end. The right-hand end is the centre-line of the slice. The model is used to investigate time-dependent response under the actions of shrinkage of the concrete and mortar, and growth of the paving. The substrate has a post-construction shrinkage strain of -0.0002. The pavers are assumed to have a long-term growth strain of 0.0002, placing them in the low growth category. It is assumed that stresses required to restrain growth are associated with the long-term modulus of the pavers. The bedding mortar is assumed to have a shrinkage strain of -0.0003. Shrinkage in the mortar grouting between pavers is -0.0005. The elastic paver-mortar modular ratio has been taken again as either $n = 2$ or $n = 20$ with the paver modulus being 15 GPa. Creep factors are 0.5 and 4.0 for paver and mortar respectively. The substrate has an elastic modulus of 27 GPa and a creep factor of 2.5. The deformed shape is presented in Figure 3 (b) for the $n = 2$ case. The construction is seen to hog but little relative movement occurs between pavers and substrate.

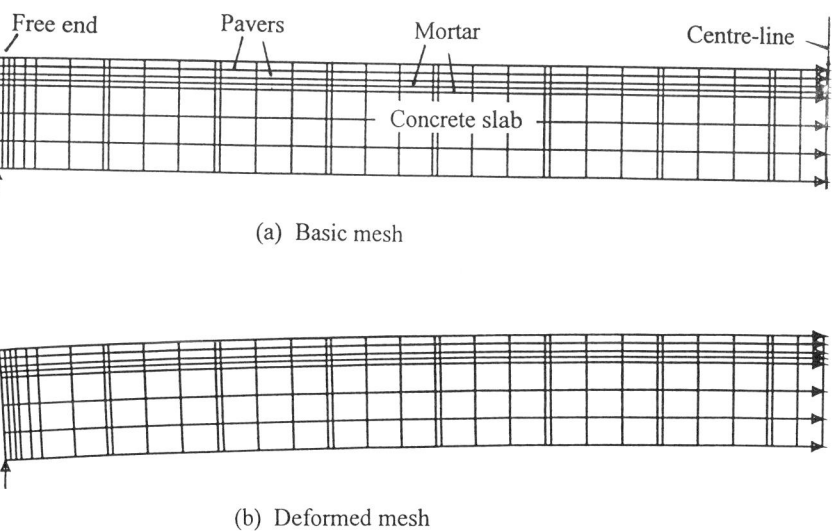

Figure 3 Computer model of paving on concrete slab, including joints

The paver-mortar interface shear stresses (Figure 4 (a)) reveal that shrinkage of the joint mortar plus the relatively high creep factor of the mortar allow the paving units to expand laterally in both directions, giving rise to saw-tooth patterns of shear strain and shear stress. For the n = 2 case, the pattern is reasonably uniform except near the free end where the highest stresses occur. For the n = 20 case, the pattern is uniform to the point where it does not feature a clear absolute maximum near the free end. Horizontal compressive stresses within the pavers (Figure 4(b)) reduce at joints but the overall patterns are similar to those in Figure 2 (b).

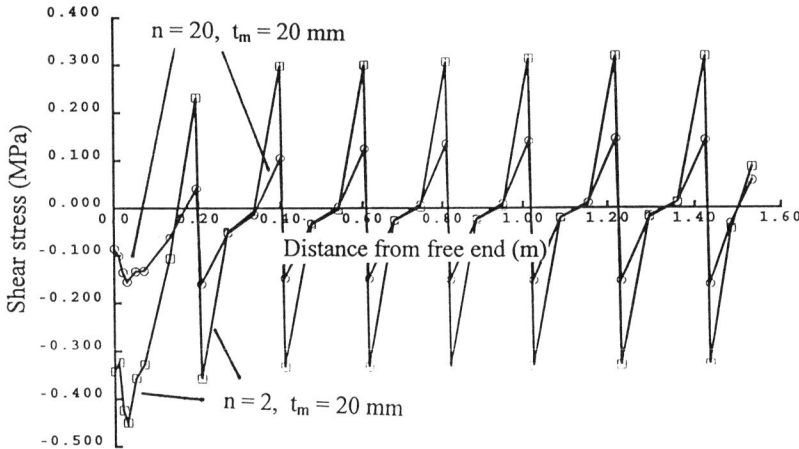

Figure 4(a) Shear stress at paver-mortar interface

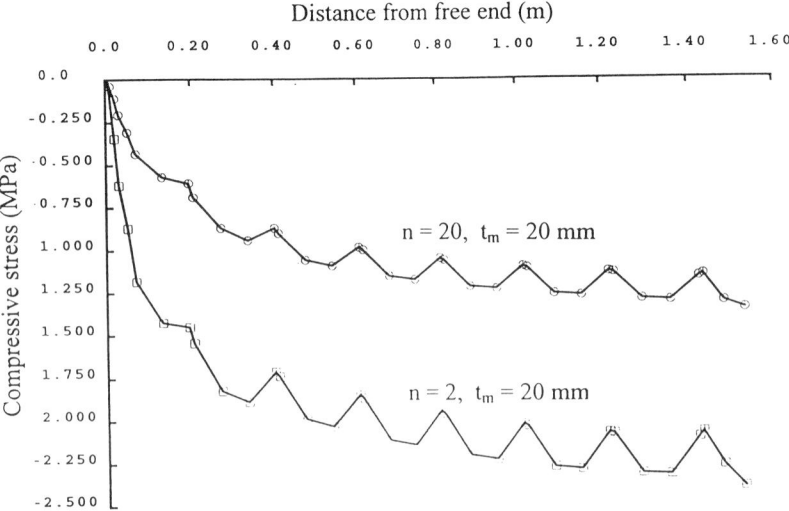

Figure 4(b) Horizontal compressive stress at mid-height of pavers

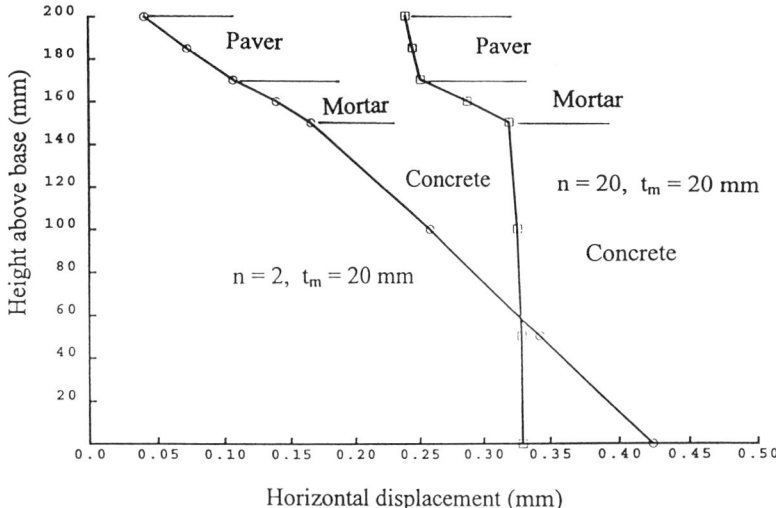

Figure 5 Horizontal displacements at free end

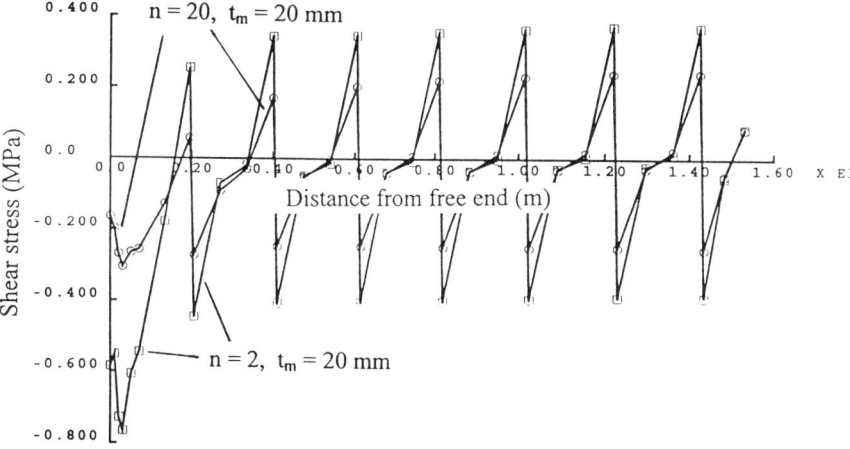

Figure 6 Interface shear stresses due to volume change and temperature differential

Figure 5 shows horizontal displacements at the free end. For the n = 2 case, these are dominated by rotation of the end cross-section as the construction hogs. For n = 20, the end displacements feature relatively uniform movements of the pavers and the substrate, and shearing deformation in the very soft mortar. The relative horizontal displacement across the mortar bed is around 12 % of the free relative movement between substrate and pavers.

Interface shear stresses due to long-term dimensional changes plus a 10°C temperature increase in the paving over the short term are shown in Figure 6. For the n = 2 case, the peak stress increases significantly near the free end. For the softer mortar bed and grout (n = 20), shear stresses increase more uniformly as units expand more freely both towards and away from the centre-line. In neither case does the shear stress (or strain) pattern feature significant accumulation with distance from the centre-line.

CONCLUSIONS

This paper has shown that the interaction between applied rigid finishes and concrete substrates is complex. Numerous conclusions are presented in the text, and these support aspects of current recommended practice. The major point of difference concerns movement joints. It is argued that these joints may be harmful, as they do not necessarily relieve accumulated strain but rather, they may introduce boundary zones of high shear stress. For this reason, movement joints should not be introduced into work that is performing satisfactorily, nor should sound work be tampered with during repair. The authors are aware of a case where the installation of a movement joint as a preventive maintenance measure led to the immediate loss of a 10 m run of paving.

A final comment is that in some instances, such as the manufacture of paving for application in aggressive environments, the quest for durability appears to have led to increased density and stiffness of surfacing units. Ironically, this leads to higher stresses in mortar layers and increased risk of failure from mechanical effects.

REFERENCES

1. STANDARDS AUSTRALIA, AS 1226.5 – 1984 Method for Determining Characteristic Expansion, Standards Australia, Sydney, 1984.

2. HENDRY, A M. Structural Masonry, MacMillan, London, 1990, 284 pp.

3. STANDARDS AUSTRALIA, AS 3600 – 1994 Concrete Structures, Standards Australia, Sydney, 1994.

4. VERMELTFOORT, A T AND WIJEN, E. The application of Electronic Speckle Pattern Interferometry for Measurement of Masonry/Mortar Stiffness, Proceedings of the Fifth International Masonry Conference, British Masonry Society, London, October 1998, pp 77 – 84.

5. MULLINS, P J AND O'CONNOR, C. Brick Shear Walls: An Experimental Investigation of the Brickwork to Concrete Interface, Proceedings of the 9th International Brick/Block Masonry Conference, Berlin, 1991, pp 317 – 324.

6. FINITE ELEMENT ANALYSIS LTD, Lusas, Finite Element Analysis Ltd, Surrey, 1998.

7. STANDARDS AUSTRALIA, AS3700 – 1998, Masonry Structures, Standards Australia, Sydney, 1998.

EFFECTS OF SHOCK VIBRATIONS ON HARDENED CONCRETE

W Zheng

A K H Kwan

K K P Lee

The University of Hong Kong

Hong Kong

ABSTRACT. In order to establish vibration control criteria for construction activities such as blasting and piling near existing concrete structures, the effects of shock vibration on concrete are studied. A laboratory testing method of subjecting concrete prisms to hammer blows, measuring the vibration intensities so induced and then determining the direct tensile and compressive strengths of the hammered prisms is specifically developed for the study. Twelve concrete mixes, of mean cube strength 25 - 50 MPa, with or without PFA added and cured at 20 or 30 °C, were tested at various ages from 12 hours to 28 days. The results reveal that shock vibration causes mainly cracking or reduction in tensile strength; but has little effect on compressive strength. The ultrasonic pulse velocity is also reduced but the amount of reduction is dependent on concrete age. From the test results, the vibration resistance of the various concrete mixes are evaluated and correlated to the other properties of concrete. Age and tensile strength are found to be the major parameters affecting the vibration resistance of concrete.

Keywords: Construction vibration, Shock resistance, Hammering test, Direct tension test, Ultrasonic test.

Mr W Zheng is a postgraduate student studying at Department of Civil Engineering, The University of Hong Kong. He earned his B.Eng. from Tongji University, Shanghai, China. His research interest includes non-destructive testing of concrete and cracking failures of concrete.

Dr A K H Kwan is a Senior Lecturer of Department of Civil Engineering, The University of Hong Kong. He obtained his doctorate from The University of Hong Kong and has acquired many years of practical experience before returning to the academia. His research topics include concrete technology, tall building structures and earthquake resistant structures.

Mr P K K Lee is a Senior Lecturer of Department of Civil Engineering, The University of Hong Kong and senior members of several professional organizations. His research covers many aspects of construction including vibration tests of structures, structural health monitoring, pile testing and foundation engineering.

INTRODUCTION

Many construction activities producing shock vibrations such as blasting and piling have to be carried out near existing concrete structures. To avoid causing damage to the nearby concrete structures, the vibrations need to be controlled. This requires knowledge of the vibration resistance of concrete. However, research on this topic has been quite limited and up to now, little is known about the effects of shock vibration on concrete.

A shock in the ground generates stress waves as in the event of an earthquake. The stress waves propagate through concrete structures in the ground and produce longitudinal strains given by the following equation:

$$\varepsilon = \frac{v}{c} \qquad (1)$$

where ε is the strain produced, v is the instantaneous particle velocity of the vibration caused by the wave and c is the propagation velocity of the wave. Hence, the effects of the shock vibration depend mainly on the peak particle velocity (ppv) of the vibration. For this reason, shock vibrations are usually controlled by setting upper limits to the ppv. It should be noted that the strain produced is dynamic and cyclic. Consequently, both compressive and tensile stresses of equal magnitude are caused by the shock vibration.

Some of the commonly adopted vibration control limits are listed in Table 1. From the tabulated ppv values, it can be seen that fairly large discrepancies exist between the various vibration control limits being applied by different authorities especially those for concrete at early ages. This is an indication that there is so far no common understanding of the vibration resistance of concrete.

Table 1 Currently adopted vibration control limits for concrete at different ages

AGE OF CONCRETE	PERMITTED PEAK PARTICLE VELOCITES AT DIFFERENT AGES, mm/s			
	Gamble and Simpson [1]	Bryson and Cooley [2]	Olofesson [3] (curing at 5 °C)	Olofesson [3] (curing at 21 °C)
12 hrs	5	5	No blasting within 30 m (10-70 hrs)	No blasting within 30 m (5-24 hrs)
18 hrs	8	5		
24 hrs	10	50		14
2 days	20	50		30
3 days	30	50	11	40
7 days	50	100	35	60
28 days	100	100	80	85
90 days	100	100	100	100

Basically, all the existing vibration control limits were established from observations in the field where the conditions are uncontrolled. Very few laboratory tests have been conducted. As described by Dowding [4], Esteves had in 1978 carried out hammering tests of concrete whereby concrete prisms were subjected to hammer blows in the longitudinal direction and the formation of any transverse cracks observed. The test results revealed that although there is a period of great susceptibility to vibration cracking between 10 to 20 hours after casting, the threshold vibration for cracks to appear during this period is as high as 150 mm/s.

As cited by Hulshizer [5], Howes had in 1979 tested the effects of shock vibration on concrete by subjecting concrete in cylinder moulds mounted on a spring supported table to hammer impact every hour during the first 7 days. It was found that even up to an intensity of 127 mm/s, the shock vibration applied had little effect on the compressive and split cylinder strengths of the concrete. Hulshizer had himself conducted recently vibration test of concrete in cylinder moulds mounted on a shaker table. The vibration tests were performed at concrete ages of 3 - 24 hours. Even though vibrations of intensity up to 400 mm/s had been applied, the concrete cylinders showed no significant change in compressive strength.

The above test results indicate that even at early age, concrete can in fact withstand fairly high intensity shock vibrations. It appears therefore that the vibration control limits in Table 1 are overly conservative and there is ample room for raising the vibration control limits so as to allow quicker and more economical construction. However, it should be borne in mind that the effects of shock vibration on concrete are not yet fully understood. It will be seen later in this paper that the major affects of shock vibration on concrete are the formation of transverse cracks and reduction in tensile strength in the longitudinal direction (the longitudinal direction is the direction of wave propagation) which have not been studied in details before.

In the present study, a new laboratory testing method in which concrete prisms are subjected to hammer impact in the longitudinal direction and then tested at the age of 28 days for their direct tensile and compressive strengths is developed. The whole experimental set up is fully digitized in the sense that all transducer signals are converted to digital form before storage and signal processing. This should yield more reliable and accurate results than those by other researchers in previous studies.

PROPOSED SHOCK VIBRATION TEST METHOD

The laboratory set up is shown in Figure 1. It consists mainly of a swing arm with a steel hammer attached at the end and a flat horizontal surface for placing the prismatic concrete specimen. The hammer is hung by a hook that can be released automatically under computer control. Roller bearings are inserted underneath the concrete specimen so that when hit the concrete specimen can move freely in the longitudinal direction. A high-g piezoelectric accelerometer is mounted on the top surface of the concrete specimen at mid-length position. The accelerometer is connected to a conditioning amplifier and then to a 12-bit A/D converter installed in a 486-PC computer. The analogue accelerometer output from the conditioning amplifier is converted to digital form at a rate of 100 kHz and then stored into the computer for subsequent data processing. The prismatic specimens are of dimensions $150 \times 150 \times 750$ mm, which are the same as those of a standard test beam specified in BS 1881: Part 118.

Its end to be hit by the hammer is attached with a 10 mm thick steel plate to ensure that the hammer impact is applied uniformly to the cross-section.

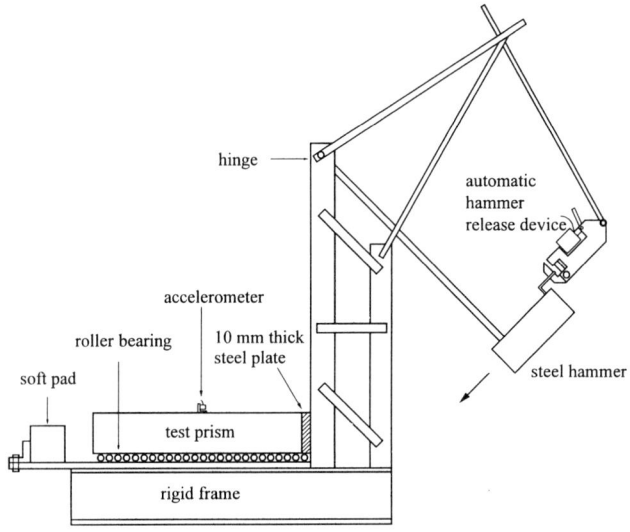

Figure 1 Sketch of shock vibration test set up

The hammer release and data acquisition operations are synchronized. When the hammer hits the end of the concrete prism, a shock vibration wave is generated which propagates along the concrete prism in the longitudinal direction. The vibration is picked up by the accelerometer which yields the particle acceleration at the point of measurement. The acceleration results are numerically integrated with respect of time to give the corresponding particle velocity, as shown in Figure 2. However, the particle velocity so obtained actually comprises of two components: the velocity of the rigid body movement of the specimen and the vibration velocity of the longitudinal shock wave. As the rigid body movement would not induce any stress, it has no effect on the concrete and should thus be removed. A trend removal process is used to remove the rigid body movement component (the trend of the particle velocity is really the rigid body movement of the specimen) so as to yield the pure vibration velocity component, as illustrated in Figure 2. It is this pure vibration velocity component that induces cyclic tensile-compressive stresses and causes cracking or damage of the concrete. From the pure vibration velocity component, the ppv of the shock wave is determined and taken as a measure of the intensity of the shock vibration.

The short-term effects of the shock vibration applied are evaluated by observing the formation of any transverse cracks on the concrete surfaces and measuring the change in ultrasonic pulse velocity of the concrete prisms in the longitudinal direction. On the other hand, the long-term effects are evaluated by continuing to cure the concrete specimens until the age of 28 days and then testing for their direct tensile and compressive strengths. The direct tensile strength is measured by gripping the two ends of the prism and pulling the prism with a tensile testing machine.

The broken pieces of the prism after the direct tensile test are then used for measuring the equivalent cube strength of the concrete in accordance with BS 1881: Part 119.

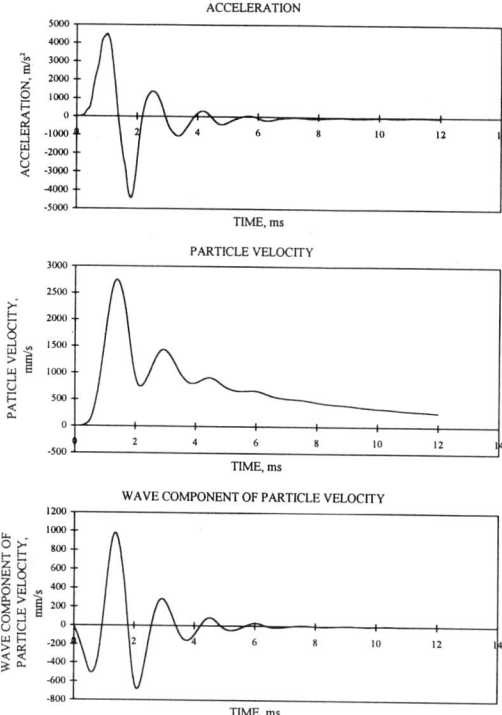

Figure 2 Obtaining particle velocity of shock wave from acceleration result

EXPERIMENTAL PROGRAMME

Twelve concrete mixes, of grades 20, 30 and 40, with no or 25 % PFA added, and cured at 20 or 30 °C were tested. The concretes were tested at ages of 12, 18, 24 hours and 3, 7, 28 days. For each concrete mix at each test age, 5 prisms (150 × 150 × 750 mm) were subjected to shock vibration test, 3 cubes (150 × 150 × 150 mm) to compression test, 2 cylinders (150 Φ × 300 mm) to split cylinder tension test and 2 cylinders (150 Φ × 300 mm) to static modulus measurement and then compression test.

Among the 5 prisms in a set, one was not subjected to any hammering to serve as a control specimen. The other 4 prisms were each subjected to shock vibrations of different intensity (the measured ppv values of the shock vibrations applied ranged from 100 to 3000 mm/s). Ultrasonic pulse velocity measurements were carried out in the longitudinal direction both before and after hammering. After the shock vibration tests, all the 5 prisms were continued to be cured at their designated curing temperature until the age of 28 days and then tested for their direct tensile and equivalent cube compressive strengths.

RESULTS AND DISCUSSIONS

The effects of the shock vibrations applied are evaluated by: (1) comparing the direct tensile and equivalent cube compressive strengths of the hammered prisms to those of un-hammered control prisms, and (2) comparing the ultrasonic pulse velocities of the concrete prisms after hammering to those before hammering.

The tensile and compressive strength results of the hammered and un-hammered control prisms indicate that whilst the shock vibrations applied have caused drastic reduction in tensile strength, they generally have little effect on the compressive strength. This is illustrated in Figure 3 where the ratio of tensile strength of hammered prism to that of control prism is plotted against the corresponding ratio for compressive strength. It is, therefore, incorrect to just consider the effect on compressive strength when evaluating the vibration resistance of concrete.

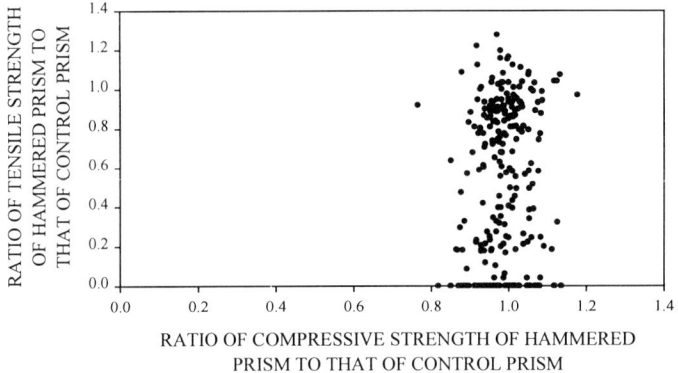

Figure 3 Effects of the shock vibrations applied on tensile and compressive strengths

Correlation of the reduction in tensile strength to the appearance of surface cracks reveal that even when surface cracks were not observed, the reduction in tensile strength could be more than 30 %. Hence, observing the appearance of surface cracks only as the criterion of vibration damage is not sufficient; there might be minor cracks formed inside the concrete causing significant reduction in tensile strength without appearing on the surface.

The change in ultrasonic pulse velocity is compared to the corresponding change in tensile strength in Figure 4. From the results plotted, it is seen that the change in ultrasonic pulse velocity corresponding to a certain percentage change in tensile strength depends on the age of the concrete. At ages of 12 hours and 24 hours, a 30 % reduction in tensile strength would be reflected by approximately 20 % and 10 % changes respectively in ultrasonic pulse velocity. At ages of 3 days or more, however, the corresponding change in ultrasonic pulse velocity becomes so small that ultrasonic pulse velocity measurement would no longer be effective for the purpose of monitoring shock vibration damage.

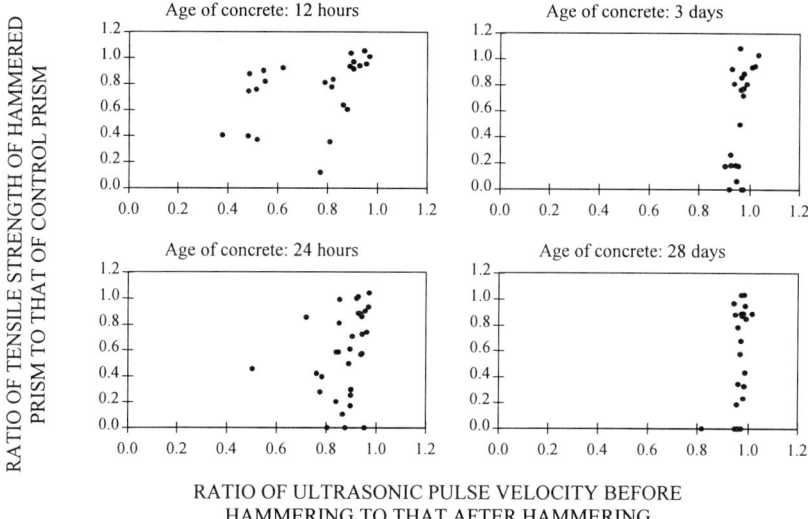

RATIO OF ULTRASONIC PULSE VELOCITY BEFORE HAMMERING TO THAT AFTER HAMMERING

Figure 4 Effectiveness of monitoring shock vibration damage by ultrasonic pulse velocity measurement

The vibration resistances of the concretes tested are evaluated as the respective vibration intensity applied (in terms of ppv) that has caused damage of the concrete. Damage, in this regard, is defined as cracking or more than 30 % reduction in tensile strength. It turns out, however, that the evaluation of vibration resistance is not straight forward, as in many case, one prism was broken into pieces when subjected to shock vibration of certain intensity while another identical prism withstood a much higher intensity vibration without damage. To overcome this difficulty, the vibration resistance is taken as the lowest vibration intensity applied that has caused cracking or more than 30 % reduction in tensile strength.

Comparing the vibration resistances of the different types of concrete tested, it is found that use of PFA up to 25 % and curing temperature within the range of 20 to 30 °C have no significant effect on vibration resistance. The major parameters affecting the vibration resistance of concrete are found to be age and tensile strength of the concrete. Treating the 12 concrete mixes tested as the same and taking their vibration resistance as the lowest vibration intensity applied that has caused any of them to be damaged, the vibration resistances of the concrete mixes at various ages are determined as listed in Table 2. The vibration resistances so obtained are plotted against the split cylinder tensile strength in Figure 5. It is seen that the vibration resistance increases with the tensile strength. Using this graph, the vibration resistance may be estimated from split cylinder tensile strength.

Table 2 Vibration resistance at different ages

AGE OF CONCRETE, days	½	¾	1	3	7	28
VIBRATION RESISTANCE IN TERMS OF PPV, mm/s	200	300	320	370	390	500

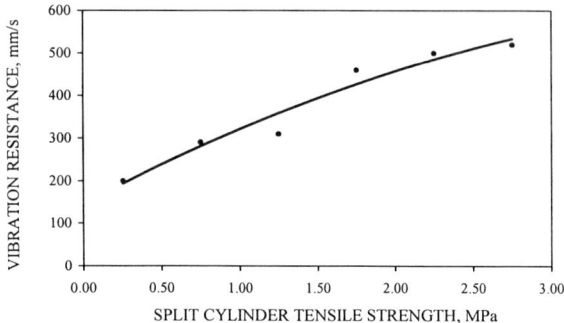

Figure 5 Vibration resistance plotted against split cylinder tensile strength

CONCLUSIONS

1. A new laboratory testing method for evaluating the effects of shock vibration on concrete has been developed. Shock vibrations up to 3000 mm/s have been applied and their effects studied.

2. Shock vibrations cause formation of transverse cracks and reduction in tensile strength in the longitudinal direction, but have little effect on compressive strength.

3. Ultrasonic pulse velocity measurement may help to monitor vibration damage of concrete only at early age and is ineffective for concrete more than 3 days old.

4. Vibration resistances of 12 typical concrete mixes have been determined. It is found that the major parameters affecting vibration resistance are age and tensile strength.

REFERENCES

1. GAMBLE, D L and SIMPSON, T A. Effects of blasting vibrations on uncured concrete foundations. Proceedings, 11th Conference on Explosives and Blasting Technique, Society of Explosive Engineering, 1985, pp 124-136.

2. BRYSON, B and COOLEY, T. Blasting procedures, Veterans Administration Medical Center, Birmingham, Alabama. Proceedings, 11th Conference on Explosives and Blasting Technique, Society of Explosive Engineering, 1985, pp 272-283.

3. OLOFESSON, S O. Applied Explosives Technology for Construction and Mining. Applex, pp 236-237.

4. DOWDING, C H. Construction Vibrations, Prentice Hall, 1996.

5. HULSHIZER, A J. Acceptable shock and vibration limits for freshly placed and maturing concrete. ACI Materials Journal, Vol.93, No.6, Nov.-Dec., 1996, pp 524-533.

ANOMOLOUS INCREASING OF STRENGTH OF CONCRETE MATERIALS UNDER CONDITIONS OF COLD HARDENING

L B Svatovskaya **T V Smirnova**
N I Uman **A M Sychova**
S G Gerke **V P Ovchinnikova**
A V Khitrov

Railway University St Petersburg

Russia

ABSTRACT. The paper discusses the possibility of anomalous acceleration of hydration and hardening processes of cement stone at an early age at –4 to 6^0C without the preliminary conditioning of products. Such acceleration is especially necessary in the northern countries where the concreting is carrying out the whole year. It was observed that the additives on the base of strong oxidizers, such as chromate or permanganate of potassium, in combination with materials with high share of covalent component of bond, are capable at an age of 7 to 14 days to raise the speed of reaction and toughness of binder materials over level reached by hardening at 20 to 22^0C. Observed anomalous strength was named a cryoactivation.

Keywords: Hydration, Low temperatures, Strength, Acceleration of hardening, Chromate of potassium, Permanganate of potassium, Phase, Pumice, Portland cement, Oxidizer.

Professor L B Svatovskaya is the Head of Department of Engineering Chemistry, Railway University, St.Petersburg, Russia. She specializes in the chemistry of the binders.

Dr T V Smirnova is a Lecturer in Department of Engineering Chemistry, Railway University, St.Petersburg, Russia. She specializes in the chemistry of the binders.

Dr N I Uman is a Lecturer in Department of Engineering Chemistry, Railway University, St.Petersburg, Russia. She specializes in the chemistry of the binders.

Mrs A M Sychova is a Postgraduate, Department of Engineering Chemistry, Railway University, St.Petersburg, Russia. She specializes in the chemistry of the binders.

Dr S G Gerke is a Lecturer in Department of Engineering Chemistry, Railway University, St.Petersburg, Russia. She specializes in the chemistry of the binders.

Dr V P Ovchinnikova is a Research associate, Department of Engineering Chemistry, Railway University, St.Petersburg, Russia. She specializes in the chemistry of the binders.

Mr A V Khitrov is a Postgraduate, Railway University, St.Petersburg, Russia. His main research interests includes the characteristics of chemical processes at low temperatures.

INTRODUCTION

The paper discusses a physical and chemical features of the state of quasisolid binder system placed at -4...-6^{0}C after mixing and molding of cement and water [1, 2].

In accordance with the classical concepts of chemical kinetics, the fall of temperature every 10^{0}C causes decrease of chemical reaction speed in 2...4 times. So at -4...-10^{0}C a speed reduction of hydration interaction of silicates and aluminates should be expected and, accordingly, the reduction of concrete strength at early age of hardening (for example, 7 days), what is really observed. Nevertheless a physical and chemical "competition" might be proposed to decelerating action of low temperatures [1]. This paper deals with the study of such kind of processes coming from basic positions listed below.

The equilibriums of exothermic reactions with reduction of temperatures are displaced toward the products of reactions. In doing so the speed of direct reaction of hydration of silicates and aluminates must rise just at low temperatures if the anions, cations or the surfaces of solids, which remove the ions Ca^{2+} or OH^- from the system into difficultly soluble compounds, are introduced into the system.

The displacement of equilibrium toward hydrosilicates and the acceleration of direct reactions of hydration must be supported by the catalysis, mainly acid-basic and (or) reduction-oxidation. From the point of view of acid-basic catalysis an introduced substance must either precipitate the ions Ca^{2+}, i.e. be a solid base and possess own electron pare, or in the other case the catalyst must be solid acid possess the capability to accept, for example, hydroxide-ions and so to rise the speed of direct reaction.

The use of the ideas of the solid-state physics on the participation of charge carriers in hydration, their transport via systems of bonds and recombination with energy evolution, which is specifically precious under freezing [2,3,4].

EXPERIMENTAL DETAILS

Materials

In this paper Portland cement M 400 and rapid-hardening cement M 500 were used. The additives of solid substances were introduced in the amount 3...5% of the mass of cement; as an additive the strong oxidizers were used – chromate and permanganate of potassium in amount 0.3...5% of cement mass and crushed pumice in amount 5% of cement mass in combination with oxidizers.

Experiment

The samples were studied in the form of pasts with a water/cement ratio equal for the control and additional, as well as in the form of concrete M 300. The mixtures were placed into freezer immediately after forming where they were hardening 28 days with strength under compression test at 7, 14 and 28 days.

RESULTS AND DISCUSSION

The kinetics of the strengthening of cement stone in conditions of hardening at low temperatures up to -6^0C (for Portland cement M 400 and rapid-hardening M 500) is shown in Figures 1 and 2. The experiments shown that at the age of 7 and 14 days the strength of material exceeds the strength of sample hardened at $20...22^0C$, i.e. the decelerating of chemical reactions and consequently the processes of hardening and strengthening is eliminated with the add of oxidizers, like K_2CrO_4, and their combinations with difficultly soluble substances of acid-basic nature, like slag pumice, in amount 3...5% of cement mass. Moreover, it can be said that an integrated additive with chromate of potassium ensures cryohydrative activation of hardening, as a result the action of low temperatures at earlier age of hardening (till 7 days) is neutralized. At an age of 28 days, as the tests shown, the strength of material exceeds the strength of control (without additives) stone too.

Derivatographic and X-ray phase analyses shown a rise of hydration grade of silicate component under freeze more than 20% and a change of phase composition of material toward the formation of low-based silicates with firmly chemically bonded water which is removed at $\approx 800^0C$. It can be supposed that in this case in conditions of quasisolid body a certain mechanism of energy transport takes place that is responsible for activation and growth of hydration grade and, correspondingly, for heat emission, the system is warming up and the following stages of hydration are activated (selfactivation).

Let's consider some positions. A binder system may be considered as quasisolid from the moment of seizing as well as the system at the temperature below 0^0C, when a cryohydration takes place (or in a general sense cryoreaction). Taking into account the ideas of the solid-state physics one of the first conclusions is the position that charge carriers may be transported via chemical bonds, including hydrogenous, with high speed, for example with aid of jump mechanism; an energy must be emitted as a result of interaction of charge carriers via certain type of recombinations. For example, it is known that the reduction of temperature promotes the discovery of specific regularities in accordance with which a new mechanisms of charge transport or interaction are possible. In such a manner low temperature polymerization became to be realized in certain conditions with unexpected high speeds (chain mechanism) or such interaction as "electron pair – lattice of solid" which result in occurrence of superconductive state of non-metallic and inorganic systems with a change of magnetic characteristics.

Supposing that under impact of low temperatures hardening system becomes to be quasisolid body, the acts of energy transmission in it may be carried out by means of quasiparticles: phonons, exitons and solitons. Besides hardening systems, as a rule, contain substances with free charge carriers – electrons or holes (silicates of different basicity when a water or alkaline solution may serve as a mixer). In hardening systems the proton, the presence of which is a consequence of occurrence in neutral or alkaline systems of activators in form of solid acids or as a mix liquid, carries out an initial act of interaction with subsequent transmission of initial product, which did not enter in reaction, towards the depth of the grains via relay mechanism and system of hydrogenous bonds of solid water. For example, if the silicate of calcium is a main substance in the system, H^+-proton reached deep grains by relay mechanism via system of H-bonds of hydrosilicates creates possible center of excitation [4,5,6].

Hereinafter the energy of excitation of centers is transmitted through the solid and is distributed in it by means of qusiparticles including exitons – hydrogen-like pairs "electron-hole".

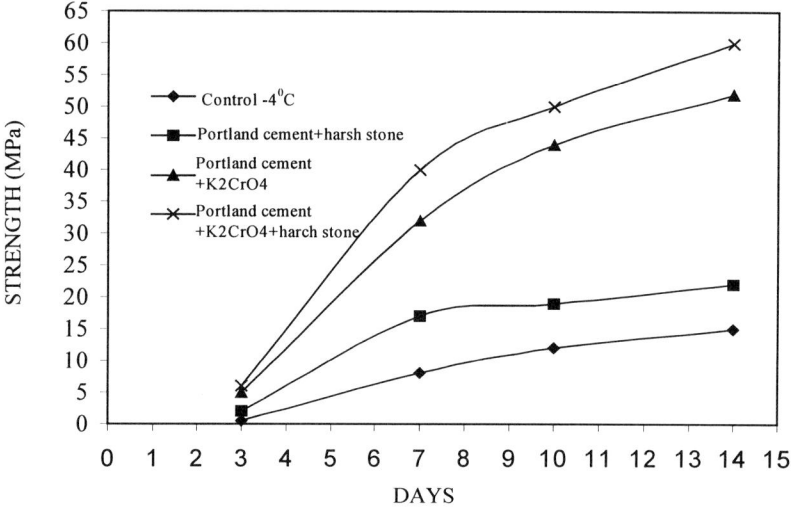

Figure 1 Kinetics of strength growth of Portland Cement M 500 at 4-6°C

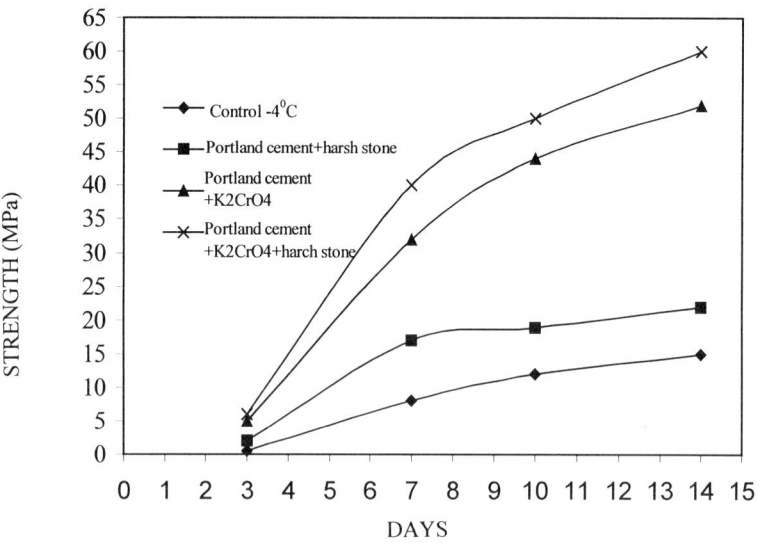

Figure 2 Kinetics of strength growth of Portland Cement M 400 at 4-6°C

Besides donor-acceptor interaction of centers the reactions of reduction of a hydrogen ion by carriers of solid phases are possible at least by those which in accordance with zone model of the structure of solid have a width of forbidden zone falls in limits till 3.5 eV [5].

The elimination of a carrier of negative charge from solid for reaction increases a number of holes, that leads to recombination with energy emission and its transmission through solid body to keep electron-hole balance.

To transform an energy of excitation to heat (i.e. to disperse an energy through a substance) a mechanism of such transformation is needed. The substances have the centers of electron adhesion – these are different admixtures, on which the energy of electron excitation is transformed into energy of heat oscillations, i.e. into phonons. Apparently a chain of energetic transformations "centers of excitation – exitons – centers of adhesion – phonons" takes place, i.e. in quasisolid system, in which solid phases with charge carriers and protons occur, there are the conditions for realization (and apparently enough active) of reactions with the development of neoformations and in certain conditions their speed is increased at low temperatures. This provide a possibility to speak about cryoreaction and cryoactivation and the Figures 1 and 2 show such processes. If molar heat capacity of threecalcium silicate very approximately consider close to value of molar heat capacity for simple substances – 26 J/mole, then even single acts of recombination with carriers, having an energy of sublevel of comparable order, can lead to rise of temperature on tens degrees at expense of heat emission, which is transmitted through the volume of solid non-conductor in form of phonons. As a result the activity may be increased by the same reason as in case of energy supply outside.

Apparently it is possible to control the cryoprocesses by means of additives of materials of corresponding electron structure, this may be also defined by zone model of solid body. These must be substances with the width of forbidden zone in limits 0...3.5 eV or dielectrics with donor-acceptor characteristics of the surface.

PRACTICAL APPLICATION OF RESULTS

In St.Petersburg Railway University cryohydration activation by additives on chromate base was used in building of university chapel. The basement of building was concreted at low temperatures. The concrete M 300 with integrated additive on base of chromate of potassium was used; the strength of concrete even at an age of 14 days corresponded to trade mark, i.e. cryohydration activation has been achieved. It is necessary to note that chromate of potassium, besides, is an additive which prevents bacterial corrosion of concrete and reinforcement.

CONCLUSIONS

1. It was discovered that some strong oxidisers like K_2CrO_4 and $KMnO_4$ in combination with difficultly soluble substances mainly with covalent bonds lead to acceleration of chemical reactions of hydration and correspondingly to hardening of concrete under freeze.

2. The influence of low temperature activators consists of an increase of hydration grade and change of phase formation toward the formation of hydrosilicates with low basicity which contain strongly chemically bound water. An explanation was proposed on the base of peculiarities of electron structure of solid, charge and energy transport and processes of recombination.

3. The concrete M 300 built in natural conditions at low temperature -12...-15^{0}C had guarantied strength at an age of 14 days.

REFERENCES

1. SVATOVSKAYA, L B, SYCHOV, M M. Activated hardening of concretes. Stroyizdat, Leningrad, 1983, pp 180.

2. SVATOVSKAYA, L B. Models of structure of solid body and hardening processes. Cement, No. 5, 1990, pp 11...12.

3. SVATOVSKAYA, L B, SMIRNOVA, T V, SYCHOV, M M, YAKHNICH I M. Activated hardening of nepheline slag. Cement, No. 7, 1989, pp 9...10.

4. SVATOVSKAYA, L B. Particularities of chemical bonds and electron structure of solid phases in production of binders and materials. In the symposium: Chemistry and technology of silicate and high-melting non-metal materials, Nauka, Leningrad, 1989, pp 252...263.

5. SVATOVSKAYA, L B. Engineering Chemistry, Part 1. Press of SPb. Railway University, St.Petersburg, 1995, pp 95

6. SVATOVSKAYA, L B. Engineering Chemistry, Part 2. Press of SPb. Railway University, St.Petersburg, 1998, pp 92

ESTIMATION OF ELASTIC MODULUS OF CONCRETE FROM ELASTIC MODULI OF ITS CONSTITUENTS AND INTERFACIAL ZONE

H Kawakami
Fukui University of Technology
Japan

ABSTRACT. This study examines the elastic modulus of concrete, with particular reference to water-cement ratio and the elastic modulus and volume fraction of aggregate. Two series of experiments are reported: 1) containing each of five rock types of gravel with a given volume fraction, 2) containing different volume fraction of gravel at two water-cement ratios. The relationships between compressive strength and elastic moduli, obtained by experiment and estimated by a two-phase model, are discussed. The elastic modulus obtained from the tests were found to be associated more closely with the volume fractions and elastic modulus of the constituents than with the compressive strength. It was also revealed that the experimental valves were lower than those calculated by the two-phase models, and that this difference increased with water-cement ratio. It is indicated that the elastic modulus of concrete is associated with the elastic modulus of constituents, their volume fraction and an inelasticity coefficient related to the cement-water ratio.

Keywords: Elastic modulus, Initial tangent modulus, Secant modulus, Two-phase geometrical model, Simplified two-phase model, Rock type, Water-cement ratio.

Professor Hideo Kawakami is Head of Architecture Course, Department of Architecture and Civil Engineering, Fukui University of Technology, Japan. He specialises in the mechanical behaviour of concrete with particular reference on the aggregate. He has written many papers on these studies. In 1994, he was awarded a prize by the Architectural Institute of Japan for his research works on the behaviour of hardened concrete. He is a member of American Concrete Institute, Japan Concrete Institute, Architectural Institute of Japan, The Society of Materials Science and Japan Union of Cement Concrete.

INTRODUCTION

Estimation of the elastic modulus of concrete has been proposed, so far, in two ways. The conventional one is empirical: estimating the elastic modulus of concrete with the compressive strength and specific gravity or rock type of gravel, as adopted in codes of ACI, AIJ and CEB-FIP [1]. The other is theoretical: predicting the average elastic modulus of concrete with the volume fractions and elastic modulus of the constituents based on the two-phase geometrical models.

The estimation formulae of these models are derived from elastic theory and the expression is more complicated as more rigorous solutions are pursued [2]. And the inelastic property caused by the aggregate interface on the elastic modulus was not taken into consideration. These would be the reason why these models found very few practical applications to concrete.

The present paper reports two series of experiment on concrete: 1) containing each of five rock types of gravel with a given volume fraction in the same mortar matrix [3], and 2) containing different volume fraction of gravel of mixed rock type in each mortar of two levels of water cement ratio [4]. The relationship between the compressive strength, the elastic modulus analysed from test results and those estimated by a simplified two-phase model proposed by the author [5] is discussed. A coefficient for the inelastic property of concrete is introduced in order to obtain the estimated elastic modulus closer to the test result. And the relationship between the coefficient and water-cement ratio is revealed.

EXPERIMENTAL DETAILS

Materials and Mix Proportions

Cement: Portland cement (ASTM Type I)

Sand: Size is under 2.5 mm, obtained from Kuzuryu River located in the central part of Japan.
Specific gravity is 2.56 in surface dry condition.
Water absorption is 2.66 %.

Gravel : Kuzuryu River gravel.

Test 1 : Pebbles are classified into 5 types of rock (Granite, Porphyrite, Rhyolite, Andesite, Sandstone).

Sieved into three groups of size (5-10, 10-15, 15-20 mm) and mixed again with same amount of each.

Test 2 : Gravel with mixed rock type were sieved into each size and mixed again in the following ratio ;
Size(mm) 5-10: 10-15: 15-20: 20-25 = 2 : 3 : 1.5 : 3.5

Table 1 Mix proportions (kg /m^3)

		W/C (%)	WATER	CEMENT	SAND	GRAVEL
Test 1	Concrete	45	168	118	253	451
	Mortar	45	308	217	465	----
Test 2	Cement : Sand = 1: 1.5 (absolute volume)					
	Water-cement ratio (%)				40	60
	Sand content in mortar (absolute volume)				0.395	0.338
	Gravel content in concrete (absolute volume) in each water cement ratio: 0.10, 0.25, 0.40.					

Test Procedure

Mortar was mixed in a pan-type mixer. The predetermined volume of mortar and prepared surface-dry gravel for each concrete specimen was hand-mixed in a bowl and cast into a cylinder mould of 10 cm diameter and 20 cm height. Three specimens were made in each mix. Capping work was done on the next day. Specimens were cured in 20°C water and tested at the age of 28 days. Longitudinal strains were measured by wire strain gauges (gauge length : 60 mm) pasted on both sides of the specimen at compression test.

ANALYSIS OF TEST RESULT

Secant elastic modulus at the stress of one third of compressive strength (E) and initial tangent modulus (Eo) are shown in Table 2 and Table 3 with the compressive strength (Fc), as well as the elastic modulus of mother rock (Ea) of each rock type of gravel. Each value of E, Eo and Fc in these tables is the mean value obtained from three specimens.

Table 2 Results of Test 1

	Ea (GPa)	Fc (MPa)	E (GPa)	Eo (GPa)	Ec (GPa)	E/Ec	Eoc (GPa)	E/Eoc
Rock Type								
Granite	57.6	39.7	30.0	35.3	34.9	0.86	38.0	0.93
Porphyrite	67.0	35.2	33.1	38.0	36.9	0.90	40.4	0.94
Rhyolite	20.8	34.7	24.4	28.0	-----	----	-----	----
Andesite	58.6	37.7	31.4	38.2	35.1	0.89	38.3	0.997
Sandstone	79.2	36.7	33.8	41.5	39.2	0.86	43.0	0.965
Mixed gravel	50.0 *	36.1	30.6	34.5	33.0	0.93	35.9	0.96
Mortar		52.6	23.9	27.6				

Elastic modulus of mother rock, estimated by a two-phase model based on the composition of volume fraction of each gravel in the aggregate. Ec, Eoc: Estimated values of E and Eo, mentioned later.)

Table 3 Results of Test 2

W/C	GRAVEL CONTENT %	Fc (MPa)	E (GPa)	Ec (GPa)	J =E/Ec	Eo(GPa)	Eoc(GPa)	J=Eo/Eoc
0.40	Mortar	50.3	23.2	-----	-----	25.6	-----	-----
	0.10	50.7	23.6	25.0	0.94	25.3	27.3	0.93
	0.25	46.6	27.0	27.9	0.97	28.3	30.1	0.94
	0.40	43.7	29.6	31.2	0.95	30.6	33.2	0.92
0.60	Mortar	34.8	19.2	-----	-----	20.5	-----	-----
	0.10	31.0	16.6	21.0	0.79	17.7	22.3	0.79
	0.25	21.9	19.0	24.0	0.79	21.8	25.3	0.86
	0.40	15.3	21.4	27.5	0.78	24.4	28.7	0.85

DISCUSSION

Elastic Modulus, Compressive Strength and Elastic Modulus of Gravel

Relationship between the compressive strength of concrete (Fc) and elastic moduli (secant modulus (E) and initial tangent modulus (Eo) are shown in Figure 1. Relationships between elastic modulus of mother rock of gravel and elastic modulus of concrete are shown in Figure 2. These figures show that E and Eo are dependent on the elastic modulus of mother rock of gravel rather than the compressive strength of concrete.

Two-Phase Geometrical Models and a Simplified Formula

Several two-phase geometrical models were proposed for estimation of elastic modulus of composite. For example, the solution of Hashin-Hansen model is summarized in (1).

$$Ec = Em [Vm\, Em + (1 + Va) Ea] / [(1 + Va) Em + Vm\, Ea] \qquad (1)$$

Ec, Em, Ea : Elastic modulus of composite, matrix, aggregate
Vm, Va : Volume fraction of matrix, aggregate --- (Vm + Va = 1)

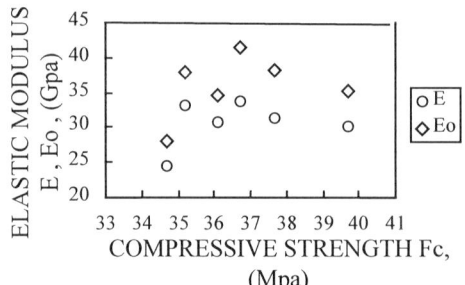

Figure 1 Compressive strength (Fc) and elastic modulus (E and Eo) of concrete

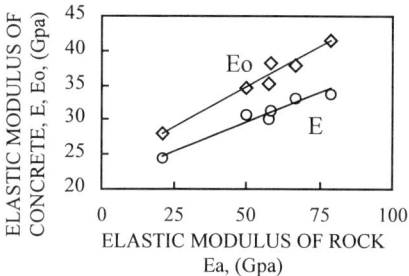

Figure 2 Elastic modulus of concrete (E, Eo) and elastic modulus of gravel rock (Ea)

The relationship of elastic modulus of composite and volume fraction of aggregate is very close to a straight line under 0.5 of volume fraction of aggregate. The author derived a simplified formula (2), substituting the rigorous solution of two-phase models as shown in Figure 3-a.

$$Ec = Em + k (Ea - Em) V \qquad (2)$$

  V : Volume fraction of aggregate, smaller than 0.5
  k : Coefficient

A coefficient k was introduced to minimize the difference between the values obtained by formula (2) and the solution of models. The coefficient k is dependent on the modular ratio of aggregate and matrix. Formula (2) is valid for every model provided with k selected for each model. Value of k for Formula (1) is shown in Figure 3-b.

Total volume fraction of gravel and sand in conventional concrete is about 0.7. However, volume fractions of sand in mortar and gravel in concrete are usually under 0.5 in each case. Formula (2) can be applied to the composite of cement paste and sand, as well as, of mortar and gravel separately with very close values to the solution of model.

Thus the two step application for mortar and for concrete each will be able to cover the range of total aggregate volume fraction larger than 0.5 and also is valid for composite containing fine and coarse aggregate of different values of elastic modulus.

The multiple application of Formula (2) would open the way for hyper composite containing several aggregates of different elastic modulus.

So far, in applying two-phase model to concrete, what the two-phase material should be, paste and total aggregate or mortar and coarse aggregate, is still in the question. The problem can be solved by the multiple application of Formula (2).

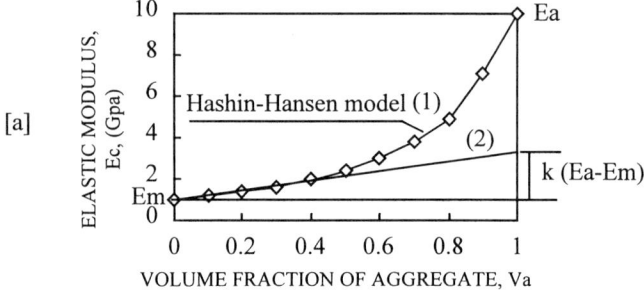

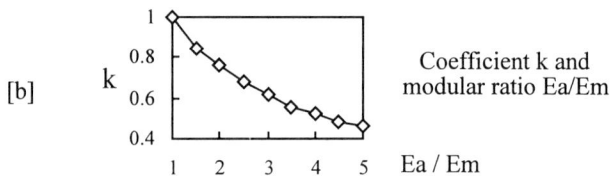

Figure 3 Two-phase geometrical model (a) and the simplified model (b)

Inelastic Property and Elastic Modulus of Concrete

Values of secant elastic modulus (Ec) and initial tangent modulus (Eoc) estimated by Formula (2) are shown in Table 3 and the relationship of them to the test result are shown in Figure 4. Test results of initial tangent modulus (Eo) are very close to the estimated values. The mean value of the ratio of test result to estimated value (Eo / Eoc) was 0.96. On the other hand, the secant elastic modulus obtained from the tests were decreased more from the estimated value. The mean value of the ratio (E / Ec) is 0.89. The lowered value of the ratios are thought to be associated with the inelastic property of concrete, as Formula (1) and Formula (2) are based on the elastic theory.

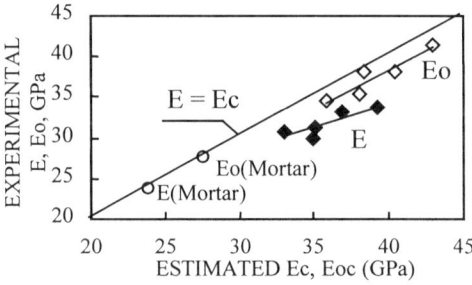

Figure 4 Elastic moduli of test result and of the estimated ---[Test 1]

Relationships between elastic modulus of test result and those estimated by Formula (2) are shown in Figure 5. Elastic modulus of test results (E) is lower in each concrete than those estimated by Formula (2).

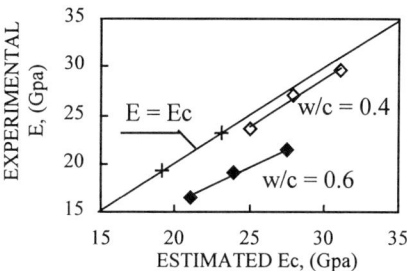

Figure 5 Elastic moduli of test result and of the estimated ---[Test 2]

A coefficient J is introduced into Formula (2) for the inelastic behavior of concrete as shown in Formula (3). The inelasticity coefficient J is defined as the ratio of elastic modulus obtained from the test to the estimated value.

$$Ec = J [Em + k (Ea - Em) V] \qquad (3)$$

J : Inelasticity coefficient

In the following discussion, the secant elastic modulus is regarded from the practical point of view. The relationship of J value for secant modulus (E) and cement-water ratio are shown in Figure 6. It is observed that J value is closely associated with cement-water ratio in a linear relation.

For practical application, Formula (3) is substituted by Formula (4), assuming the k value as 0.67 (the mean value for the practical range of modular ratio of 2 - 3.5)

$$Ec = J [Em + 0.67 (Ea - Em) V] \qquad (4)$$

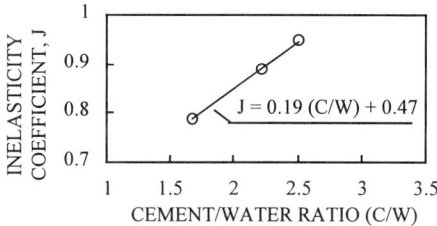

Figure 6 Inelasticity coefficient J and cement-water ratio

Figure 7 shows the relationship of test results and those estimated by Formula (5). Secant elastic moduli estimated by Formula (5) are found to be close to the test result.

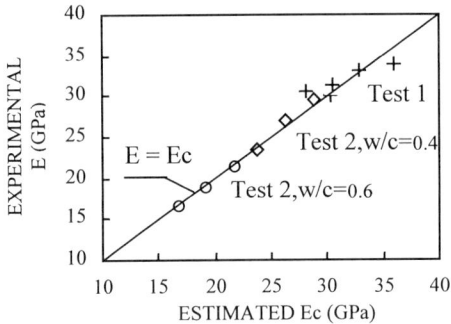

Figure 7 Elastic modulus tested and the estimated by Formula (4)

CONCLUSION

Elastic modulus of concrete is associated more closely with the elastic modulus and volume fraction of constituents, rather than the compressive strength of concrete. The two-phase model and the simplified model proposed by the author were preferable for estimation of elastic modulus of concrete. But the estimated values thus obtained were higher than the values from experimental tests.

By adopting a coefficient, J, representing the inelasticity of concrete, it was found that the simplified two-phase model gave values close to the test results, suggesting that it would be an approach to the practical estimation of elastic modulus of concrete. For the concretes considered, the J value was linearly related to the cement-water ratio. Experimental data regarding the inelasticity coefficient J for wider range of mix proportions are to be compiled.

ACKNOWLEDGEMENTS

The author would like to express his appreciation to Prof. Shizuka MIURA in Fukui University for his advice on the rock and also to Mr. Keiichi WAKI (technician) and students for their cooperation in experiments and analysis.

REFERENCES

1. AIJ Structural Committee. Young's modulus applied for rigidity evaluation of concrete structure J of Architecture and Building Science, Vol.100, No.1241, 1985, pp.36-47.

2. HANSEN,T.C. Theories of multi-phase materials applied to concrete, cement mortar and cement paste. "The Structure of Concrete," Proceedings of an International Conference, London, September 1965

3. KAWAKAMI, H. Effect of aggregate type on the mechanical behavior of concrete. Proceedings of the RILEM International Conference on Interfaces in Cementitious Composites, Toulouse, 1992, pp. 179-186.

4. KAWAKAMI, H. Multiple application of a simplified two-phase model for estimation of elastic modulus of concrete. Transactions of Japan Concrete Institute, Vol.19, 1997, pp. 33-40.

5. KAWAKAMI, H. Approximation of two-phase structural models for evaluation of Young's modulus of concrete. Proceedings of the 21st Congress of Japan Union of Cement Concrete, 1994, pp.85-88.

INDEX OF AUTHORS

Aba, M	79-88		Svatovskaya, L B	203-208
Achouche, J	163-170		Sychova, A M	203-208
Amleh, L	27-38		Toman, J	63-70
Aoun, H	163-170		Uman, N I	203-208
Bărbuță, M	171-176		van Breugel, K	121-132
Beeldens, A	177-184		van Gemert, D	177-184
Bijen, J M J M	17-26		Zheng, W	195-202
Bílek, V	71-78			
Bournazel, J-P	111-120			
Buckles, G J	153-162			
Buyle-Bodin, F	163-170			
Černý, R	63-70			
Czarneki, L	177-184			
de Almeida, I R	133-142			
de Rooij, M R	17-26			
de Souza, R H F	133-142			
Demura, K	177-184			
Drchalová, J	63-70			
Dux, P	185-194			
Fairhurst, D	51-62			
Felício, M D	133-142			
Gacel, J-N	111-120			
Gerke, S G	203-208			
Heywood, B R	153-162			
Hošková, S	63-70			
Igusa, T	1-16			
Jurek, P	63-70			
Kawakami, H	209-218			
Kendall, K	153-162			
Keršner, Z	71-78			
Khitrov, V	203-208			
Klečka, T	63-70			
Koenders, E A B	121-132			
Kwan, K H A	195-202			
Lee K K P	195-202			
Leong, T-W	143-152			
Milnes, K	153-162			
Mirza, M S	27-38			
Mullins, P	185-194			
Ohama, Y	177-184			
Onet, T	171-176			
Otsuka, K	79-88			
Ovchinnikova, V P	203-208			
Pettersson, K	89-96			
Platten, A K	51-62			
Schwartzentruber, A	111-120			
Shah, S P	1-16			
Shaw, I M	39-50			
Short, N R	39-50			
Shoya, M	79-88			
Smirnova, T V	203-208			
Subramanian, S	97-110			

SUBJECT INDEX

This index has been compiled from the keywords assigned to the papers, edited and extended as appropriate. The page references are to the first page of the relevant paper.

Abrasion 97
Accelerated
    corrosion 27
    hardening 203
    testing 1
Additions 17
Adherence 177
Adhesive failure 185
Admixtures 17, 133
Aggregate 121, 153
Analysis 39
    structural 143
Anchorage 163
    length 133

Beam 163
Bond 133, 163, 171
    deterioration 27
    stress 171

Cast-in-situ composite member 143
Cement 17, 39
Cement-based composite 71
Characterisation 153
Chloride ion profile 27
Chromate of potassium 203
Claddings 185
Coagulation 17
Composition 153
Concrete 121, 163
    performance 97
    precast 143
    reinforced 27
    repair 89
Condensed silica fume 133, 171
Construction vibration 195
Corrosion 27, 89
Cracks in concrete 143
    longitudinal 27
    transverse 27
Creep 143
Curing conditions 177
    steam 79
Cyclic loading 163

Debonding 185

Deep-drawing tools 111
Differential temperature 143
Diffusivity 51
Direct tension test 195
Durability 1, 39, 89, 97
Effective fracture toughness 71
Elastic modulus 209
Electrochemical testing 27
Engineer 97
Experimental design 1

Facades 185
Failure 185
Flocculation 17
Floors 185
Fly ash 133
Fracture characteristics 71
Frost resistance 79

Hammering test 195
High Performance Concrete (HPC) 111
High strength concrete 133
Hydration 121, 203

Impact 97
Initial tangent modulus 209
Interactions 39
Interface 17, 39
    physical 97
    societal 97
Interfacial zone 121

Lollipop specimens 27
Low temperatures 203

Mechanical load 63
Microcracks 79
Microstructure 71, 121, 153
Moisture
    diffusivity 63
    transport 121
Monotonous loading 163

Natural pozzolan 133
Non-destructive evaluation 1
Numerical simulation 121

Oxidizer   203

Particle-packing   17
Paste   121
Pavers   185
Permanganate of potassium   203
Permeability   79
Phase   203
Polymer-cement ratio   177
Polymers   39
Porosity   51
Porous concrete   177
Portland cement   153, 203
Pull-out bar   171
Pulverized-fuel ash (PFA) 27
Pumice   203

Random variable   71
Reinforcement   163, 171
       steel bars   133
Relaxation of stress   143
Reliability engineering   1
Rigid finishes   185
Rock type   209

Scientist   97
Secant modulus   209
Service life prediction   1
Shock resistance   195
Shrinkage   143
Silica fume   51
Simplified two-phase model   209
Simulation   71
Specifier   97
Statistics   1
Strength   97, 121, 203
       tensile   143
Superplasticizers   111, 133
Supplier   97
Surface   39
       state   111
       tension   111
Suspension   17
Syneresis   17

Testing
       electrochemical   27
       ultrasonic   195
Tiles   185
Time dependent stresses and strains   143
Toughening mechanism   71
Transition   17
       zones   89
Two-phase geometrical model   209

Vapour transfer   51

Water
       cement ratio   209
       sorption   51
       vapour permeability   63

X-ray inspection technique   79